BRITISH WOODLAND

Also by Ray Mears

We Are Nature

BRITISH WOODLAND

HOW TO EXPLORE THE SECRET WORLD OF OUR TREES

RAY MEARS

EBURY
SPOTLIGHT

Spotlight, an imprint of Ebury Publishing
20 Vauxhall Bridge Road
London SW1V 2SA

Spotlight is part of the Penguin Random House group of companies
whose addresses can be found at global.penguinrandomhouse.com

Penguin
Random House
UK

First published by Spotlight in 2023

www.penguin.co.uk

A CIP catalogue record for this book is available from the British Library

ISBN 9781529109993

Printed and bound in Great Britain by Clays Ltd, Elcograf S.p.A.

The authorised representative in the EEA is Penguin Random House
Ireland, Morrison Chambers, 32 Nassau Street, Dublin D02 YH68.

MIX
Paper from
responsible sources
FSC® C018179

Penguin Random House is committed to a sustainable future for
our business, our readers and our planet. This book is made from
Forest Stewardship Council® certified paper.

To Ruth

Darling, gorgeous wife,
I love that you love trees as much as I do.

You are as generous as the Birch,
as adaptable as the Willow,
and as constant as the Oak.
Our love grows ever stronger,
like the Yew.

Thank you for your patience, your love,
and for always being there for me.
I love you.

CONTENTS

1

AWAKENING

After the retreat of the last glaciers to cover Britain, herds of reindeer would have been one of the animals that lured hunters onto British soil in the late Upper Palaeolithic.

Walking in British woodland today, with its diverse range of tree species that reflect the mild and gentle nature of our temperate climate, it is difficult to appreciate that, for the past two and a half million years, Britain has mostly been in the grip of the Quaternary Ice Age, the most recent of a succession of at least five earlier ice ages. During this time, our climate has been dominated by a succession of glaciations: periods of extreme cold which have been interrupted by warmer periods, known as interstadials. It is only during these interstadials that more temperate conditions have prevailed. Even though these episodes may last for thousands of years, they are but brief interludes in the more normal frigid landscape when observed from the much longer perspective of a geological timescale. Even today, were it not for the warmth afforded us by the Gulf Stream, we would be an island of a subarctic nature. One quick glance at a globe and it becomes apparent that London shares its latitude with more frigid locations such as Lake Mistassini, James Bay, Lake Winnipeg, Banff National Park and Lake Baykal.

Geologists have established that the most recent glaciation in Britain, the Devensian, held us in its icy grip for 100,000 years. It is hard to imagine but 22,000 years ago, at the time of the glacial maximum, Scotland, Wales, most of Ireland, Northern England and much of the Midlands was covered by a glacial sheet of ice one kilometre thick. The enormous power of those glaciers carved, sculpted, moulded and eroded the ground, remodelling the landscape into the Britain that we so cherish today.

It was around 15,000 years ago that the temperature began to rise year on year, gradually releasing Britain from its glacial mantle. Although Britain had been an island separated from Europe in earlier warm periods, at this time, as the land began to blush with returning vegetation, the outline of Britain was unrecognisable, consumed within the vast northwest corner of Europe; land that extended from northern Brittany around the west of Ireland, encompassing the Outer Hebrides and Shetland islands to northern Denmark. Where the North Sea is today was a great expanse of rolling hills, valleys, rivers and marshland that we call Doggerland.

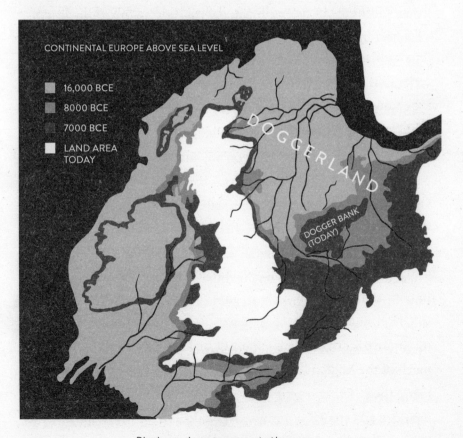

CONTINENTAL EUROPE ABOVE SEA LEVEL

- 16,000 BCE
- 8000 BCE
- 7000 BCE
- LAND AREA TODAY

DOGGERLAND

DOGGER BANK (TODAY)

Black area here represents the sea

Instead of what is now the English Channel, the mighty Channel River flowed west through a broad valley into the Bay of Biscay, fed by many tributary rivers which included the Solent, the Rother, the Thames, the Seine and the Rhine. Connected in this way, so long as they could cope with the climate, animals, plants, trees and people could easily find their way onto British soil, particularly during cold winters when rivers were frozen and could be walked across. We know now that that our climate oscillates from warm to cold under the influence of solar radiation in concord with our Earth's orbit and the tilt of its rotational axis, with warm interglacial periods coinciding with periods of peak solar radiation. It is highly likely that, mocking our recent Brexit, the cold will return in the future to connect us once again to Europe. But don't rush out to purchase an arctic parka just yet, we are only a short way into this warm period. The next cold phase is not predicted for another 50,000 years. It is also not yet known to what degree the effect of reducing solar radiation will be outweighed by our malign influence, which is causing the Earth to warm faster than on any previous occasion.

While geologically cold phases tend to develop slowly and end abruptly, for the plants and animals, including people that experienced the most recent change from an arctic to a temperate climate during the last few thousand years of the ice age, the change was mercurial. Research has revealed that the warming was punctuated by three dramatic returns to a cold climate. It was as if the cold was toying with us, like a snow leopard tormenting its prey, only reluctantly relinquishing its dominance over the land.

We can identify the first significant period of warming, the Windermere Interstadial, around 14,700 BP (before the present time). That's when the climate warmed from the frigid arctic climate to one that is described as 'cold temperate'. In effect the summers were still cool and the winters mostly hovered just below freezing. Even before the weather had significantly warmed, sedges, mosses, grasses and sorrels seized the opportunity to colonise the barren ground. Along with the first trees – the diminutive dwarf willow (*Salix herbacea*) and dwarf birch (*Betula nana*) – a juniper (*Juniperus communis*) may have been seen here and there and, in sheltered south-facing valleys, perhaps even the odd weather-beaten downy birch tree grew. But this was a windswept tundra, a moorland landscape mostly devoid of trees, where wood was scarce.

As a wildlife filmmaker, I wish that I could have filmed this landscape. I have often pictured the view from the Chilterns. In the distance, the bulky form of mammoths grazing on grasses and the forest of tiny dwarf willow trees; for safety we stay down wind of them. Perhaps their rumbling murmur of content is audible on the breeze, although their days are numbered as, with warming conditions, they will lose their niche in the ecosystem as woodland replaces grassland. There is a high probability that the calm scene will be interrupted by wild horses, the stark whinnying of two frenzied stallions, teeth bared and mouths foaming, fighting as they compete for dominance over a harem of mares. Dodging the unpredictable stallions, we see in the middle distance some aurochs (the extinct wild ox that is the origin of modern cattle), quietly grazing in the valley among some wispy trees while, there on the moorland above, we catch sight of a herd of reindeer, a reminder of the continuing cold nature of the climate and the harshness of winter. They are alert, the fur on their shoulders ruffling in the chill breeze; all are looking in the same direction, noses to the wind. Just below them we detect the loping trot of a wolf pack, relentlessly searching for an opportunity to circle round downwind of the herd. They – along with the brown bear, the lynx, the red fox and the cheeky arctic fox – are assuming the mantle of predators from the cave lion and cave hyena that are becoming rarer year by year, if not already extinct.

While we survey the scene we are attracted to a human laugh; two young women wearing well-tailored clothing of skins are passing by. One is carrying an arctic hare; they have been checking snares. These are our direct ancestors, *Homo sapiens*, identical to us. We cannot understand their words, but their mood is cheerful and easy to read. If we were to follow them, we would find them returning to a cave in a limestone gorge that today we know as the Cheddar Gorge. These are a confident people who have walked here from southern Europe. Their culture is centred on hunting wild horses and reindeer. They make distinctive flint tools, blades struck from a core, blunted on one side and with angular truncated ends, rather like a modern Stanley knife blade. The uniqueness of these tools was first recognised by archaeologists excavating a cave in South West France called Abri de la Madeleine; ever since, these people have been called Magdalenian. They are a mysterious people, who live at a time of great cold which to the modern mind suggests struggle, and yet they will be best known for the incredible art they are to leave behind, particularly cave paintings at La Grotte Chauvet and Lascaux in France or at Altamira in Spain. There, using the walls of their caves as dynamic three-dimensional canvases, they graphically depicted the animals of their world. These are not just any rock art; they are true artistic masterpieces. With deft and subtle use of ochre and charcoal, to my mind the equal of the greatest Renaissance artists, they reveal their intimate familiarity with animal anatomy but, more importantly, their feeling, respect and spiritual engagement with the animals of their world. We can only imagine the magic of observing these possibly religious frescoes by the flickering light of a lantern fuelled by fat, when the images would have triggered the heightened awareness of these late Palaeolithic hunters, the animals seeming to manifest themselves, to come alive and to leap from the very walls. In Britain, too, rock art has been found in Nottinghamshire at Creswell Crags: and engravings of bison, ibis and other forms have only recently been found on cave walls. It was not just rock art that these people left behind; they also loved to engrave, carve and embellish their tools made from mammoth ivory or antler. At Creswell Crags a beautiful engraving of a horse made on a rib bone highlighted with ochre was found. Surely, they must also have carved wood – if only it had survived what marvels might we observe.

In 1866 at the Montastruc rock shelter, near Montauban in Southern France, the carved tip of a mammoth tusk was found. The carving depicts two reindeer: they have long been described as swimming reindeer, but I believe this interpretation to be incorrect. If you have spent time with reindeer herders in late September, you will immediately recognise that the carving depicts the courtship behaviour of mating reindeer, when the bull in a hormonal frenzy chases the cows, lowering his head to sniff their behinds. The carving not only beautifully depicts this event in the wild calendar, it also captures the primal energy of the moment and conveys its portent: an event that would have symbolised continuation for a people who depended upon reindeer for their food, clothing and tools. Having travelled in the High Arctic wearing Inuit clothing fashioned from caribou fur, I can fully appreciate the sentiment of the artist. In a cold land with few trees, the reindeer with their hollow insulating hairs made it possible to thrive in the coldest weather. To my mind these people were real specialists in cold-weather living.

Here in Britain these Magdalenian hunters left traces of their presence widely across the country. They were clearly confident, able travellers, perhaps exploiting large territories. They used spears tipped with stone blades. Broken spear points have been found in cave sites, testimony to time spent replacing and repairing their equipment. From the antlers, tusks and bones of their prey they fashioned harpoon points, spear throwers and fine sewing needles. Following fire, the second greatest human innovation was the needle (60,000 BP), enabling warm clothing to be made to allow tropically evolved humans to populate the northernmost reaches of the Earth. Tendons, dried and separated, provided the sewing thread which was strong enough to often outlast the leather garments. Although artists have habitually illustrated these people wearing crudely fashioned clothing, one glance at their carving or artistry is all that is required to suggest that their clothing would also have been well tailored and perhaps even fashionable. All of my friends in indigenous northern communities, be they Inuit, Siberian Evenk or Sami, are inheritors of a long tradition of making fur clothing which is practical, beautifully tailored and artfully designed. Their garments are frequently made from many small pieces of contrastingly coloured fur for artistic effect which often serves to define their cultural identity. In recent years I have seen Inuit seamstresses realising traditional designs in contrasting colours of fleece, the fur of the modern age.

Intriguingly it is believed that these people had begun the process of domesticating wolves, perhaps using them to assist in hunting or as pack animals. A wolf pup bone found in Gough's Cave in the Cheddar Gorge may represent this while, in Europe, bones from domestic dogs have been found from later Magdalenian sites on the shore of Lake Neuchâtel, Switzerland.

From the Magdalenian rubbish heaps, archaeologists have been able to investigate the bones of their prey, the parallel cut marks left by the wavy edge of their stone blades revealing the way in which they skinned the animals, how the tendons were carefully removed and how the joints were disarticulated. The bones were cooked, de-fleshed and smashed to extract the tasty and energy-rich marrow which could have been eaten immediately or saved for inclusion with dried meat as pemmican or travelling food. Alarmingly, at Gough's Cave such marks were also found on human bones along with evidence of the bones being gnawed clean by human teeth. A cranium had been painstakingly converted into a cup and one forearm had been adorned with deliberate geometric zig-zag patterns before being broken and the marrow extracted. Given the seeming abundance of food, it is thought that this cannibalism was ritualistic. But who was eaten and why? Questions that will remain a secret from the end of the age of ice to the end of time.

I have long wondered about these people and their knowledge of trees and plants. Judging by their obvious knowledge of animals, bone and antler, they had a sophisticated understanding of natural resources. They needed wood for spears and many other tools that remain invisible to us, but might they have managed life with very little wood? Although the region was not devoid of trees, it is believed that they lived at a time when wood was scarce. Does their extensive use of caves hint at this? Did they perhaps live a lifestyle akin to that of the traditional Inuit? They certainly had the means to cope with a cold tundra environment. In Europe, Magdalenian tent sites have been identified where they appear to have constructed fires from small-diameter willow and birch brushwood, burning this under a covered hearth of rocks. One explanation for this is that such hearths may have reduced the need for fuel and extended the duration of the fire's warmth as heat continued to radiate from the heated rocks. A clue is perhaps also to be found in their use of fat lamps. In Europe, nearly 300 Magdalenian fat lanterns have been found, some associated with 'art galleries' deep inside caverns, but intriguingly the majority have been found at open-air sites. While these have usually been regarded as sources of light, were they perhaps also using them as portable stoves for travel in a treeless land? If so, the first tree of significance to humans to grow after the ice age would likely have been the dwarf willow, the world's smallest tree, which only grows to six centimetres in height, today mostly found high on the mountains of northwest Scotland.

In 1996, I was shown by Ham Kadloo, an Inuit elder from Pond Inlet, Baffin Island, how to make a *qulliq* or *kudliq*, the traditional Inuit fat lantern. Historically carved from soapstone, the qulliq is a D-shaped shallow tray, with its deepest point in the crook of its curve. A wick is laid along the length of the straight side; it absorbs the liqui-fied fat and burns brightly. The warmth from the wick gently melts the fat in the tray maintaining a constant supply of liquid fat to the wick. The wick is critical to the device; it is made from a mixture of dried moss (*Maniq*) and seed down from the Arctic willow (*Suputit*). It is the seed component that gives the wick its easy combustibility. While cotton grass down (*Pualunnguat*) can be used as a substitute, the Arctic willow is considered the best wick material. An Inuit lady explained to me that managing a qulliq is a special skill, and that it was essential to always clean the down before use, carefully removing the small seeds that would otherwise impair the efficiency of the wick.

Demonstrating both Inuit practical adaptability and craftsman-ship, Ham hammered his qulliq from a piece of waste aluminium. 'Lighter, more compact, does not break. Better for hunting,' he explained, also pointing out that old car hubcaps are perfect for the job. In operation, though, everything else was traditional: seal fat was first pulped with a caribou antler hammer, to liquify the fat and be placed in the qulliq, then a small heap of the wick material was laid along the straight edge of the lantern. This was gently encouraged

Qullic

to absorb some of the liquid fat and the stove was ignited with a Zippo lighter. In the past, he explained, the stove was lit with a flint and steel using suputit as the tinder. I learned that day that there is a special art to maintaining the wick so that it burns just so, with an even flame of equal height and intensity along its entire length. It was warm enough to keep the fat melting but not so intense that the precious fat was consumed too quickly, and certainly not smoking with unpleasant sooty fumes.

Despite the modern use of Coleman petrol stoves, the qulliq remains close to the Inuit's heart, for it is more than a lantern: it is central to their culture, the traditional source of warmth in an igloo, essential light during the long winter nights to repair and maintain clothing, a stove to melt ice, to boil water and to cook over and, in an emergency, to dry wet clothing on a drying frame, or *ittaq*. In the Inuit household the qulliq was made by the father but was the domain of his wife, who lovingly trimmed the wick with a small, hooked stick, the *taqquti.* The perfectly trimmed flame of the qulliq was a visual mani-festation of a woman's strength, representing her devotion, love and care for her family, but most importantly the qulliq was and still is a symbol of human survival, just as the campfire is in a remote forest.

The dwarf willow and the arctic willow can be used identically. Did the down of dwarf willow serve the same function for those Magdalenian hunting parties? Did they have to gather their winter supply in the summer? Did they eat the leaves of dwarf willow like the arctic willow was eaten by the Inuit, to temper the taste of rancid fat? Did they peel the roots and eat them to cure sore throats? Sadly, we shall never know. But there is a distinct possibility that they employed this same technology. (Incidentally, a significant number of Magdalenian fat lamps seem to have been ritually broken. Does this perhaps represent the passing of the owner?)

However, I think that there was another tree that, even more than the dwarf willow, would have emboldened the hearts of those early people when they considered pursuing horse herds into the vast trackless wilderness of tundra Britain.

The Magdalenian hunters exploited Britain's hunting ground for around 600 years. When travelling between caves and rock overhangs, they must have employed tents. What design were they and how did they transport them? If they had been here longer, perhaps enough of their wooden artefacts might have survived for researchers to study today. But around 14,100 BP the climate took a downturn in temper-ature that forced them and the game they depended upon further south to warmer climes in Europe. For 200 years, Britain was once again an arctic desert during the period known as the Older Dryas.

At 13,900 BP, the climate once again thawed to a cold temperate ecosystem. Grasses and dwarf willow grew again but on this occasion the pollen record reveals that they declined, while birch and juniper proliferated. This change from open grassland to light woodland predominantly comprised of birch, juniper and willow did not favour the mammoth which now disappears from our fossil record and the herds of wild horses also seemed to have dwindled in the more arboreal landscape. By contrast, woodland-loving red deer, giant deer and aurochs proliferated and were joined by elk. Exploiting these resources came a different human community, today called the Federmesser, after the German word for a penknife, which aptly describes the characteristically shaped flint tool used to identify these people. This tool has a curved, blunted back, like a small penknife blade. The trail of these blades suggests that these were a different people, related to groups living in Northern Europe. To arrive here they would have walked across the vast hunting grounds of Doggerland. They may have camped on the expanse of gravelly hills that are now the rich fishing ground of the North Sea, called Dogger Bank. For centuries, trawlermen fishing the bank have hauled up Federmesser harpoons in their nets, along with other treasures of the past.

The Federmesser did not rely on caves to anything like the same degree as the Magdalenians: most of their sites are found in the open, which makes the detection of their traces challenging for field archaeologists. Many have undoubtedly been swept away by the plough and other agricultural or industrial activities. Most of the sites discovered in Britain are found in the south and the east, perhaps supporting the notion that Britain was the far reach of their territory, although given the sparsity of finds and the age of the period they may well have been travelling anywhere on land we now call British. Today, oil prospecting and aggregate extraction is enabling archaeologists to better understand the significance of Doggerland and even to search likely areas of the seabed for evidence of past ecosystems and people. We shall encounter some of their extraordinary finds later. Intriguingly a bison bone from the time of the Federmesser was recovered from Doggerland, engraved with a zig-zag pattern very similar to that made on the human bone in Gough's Cave. Does it represent a long-continuing spiritual belief? Many theories have been postulated about the meaning of the engraving; I have often wondered if the pattern represents the northern lights, perhaps the perceived spiritual after-world. While it is fun to speculate, I am conscious that every indigenous community I have had the privilege to work with maintains a tradition of according great respect to the spirits of their departed prey.

Apart from their stone tools, little organic material survives, but we do know that they fashioned characteristic slender antler harpoons with barbs neatly carved along one or both edges. These were used for hunting game and fish. They also travelled with domestic dogs, though whether these were beasts of burden, hunting dogs or a food source is not known. At Rookery Farm, Great Wilbraham in Cambridgeshire, a Federmesser site was found in 2002, located on a slight slope overlooking a flat expanse of land close to two springs. The site suggests a very short period of occupation, possibly a tented encampment or a bivouac. Proximity to water is obviously advantageous, but its location is intriguing. It is often suggested that this slope was beneficial for spying distant game, which could certainly be the case. However, there can be other reasons for choosing such a location. When I bivouac in this type of landscape and climate, I search for a campsite close to water and to firewood, which is slightly elevated, without being unnecessarily exposed. In cold months the elevated ground avoids the chill of low-lying ground which fills with an invisible river of cold, damp air, particularly at night. In summer I also choose slightly elevated ground, but this time to gain a breeze to reduce the unwanted attention of biting insects. I do not imagine that the decision making was any different all those years ago.

The light footprint of the Federmesser on the land partly results from their use of tents. Living in a period with more trees, they could heat their tents in winter with ample firewood. A well-heated tent is far warmer than a cave. These tents might have resembled any of the nomadic tents from the arctic, but if I were to guess which design they might have used it would be the most employed arctic tent design, the *chum* or *lavvu*, similar to those made by Siberian and Lapland communities. Picture a cone-shaped construction of slender birch poles covered with deer skin hides, possibly with the fur left on in winter and without fur in the summer. In summer it might also have been possible for small hunting parties to simply thatch a temporary shelter with bushy young birch saplings. With wood now more abundant, it is likely that they left the pole frames standing when moving on, taking only the tent covering with them, the poles remaining for future use and visible from a distance indicating the location of the campsite. Although these groups bivouacked in Britain for a thousand years, only scant evidence of their presence has so far been found. They seem to have been true masters of no-trace camping.

At 12,900 BP, Britain was suddenly plunged into polar desert conditions, by an abrupt drop in temperature that occurred in less than 50 years. The Loch Lomond Stadial, or Younger Dryas, would chill Britain to its core for the next thousand years, before ending as abruptly as it began. (The abruptness of the climatic changes of this stadial are today of great interest to climatologists studying the current trends in the global climate.) The Younger Dryas seems to draw a line under the Federmesser culture, the assumption being that the climate became too hostile to support human existence across much of Northern Europe at that time. Perhaps though, as the Younger Dryas began to recede and the climate showed early signs of improving, some adventurous parties of people once again ventured north.

Although scarce, there are some tantalising finds from this period: humanly modified bones and stone tools that are fashioned in a new way, with tanged points and long blades. Necessity being the mother of invention, it seems likely that the cold of the Younger Dryas was the genesis for a raft of technological advancement, of which we glimpse only the most durable manifestations. Did people improve their use of skin clothing, refine their tailoring skills, improve their mobility and navigational knowhow? We can only guess at their new abilities and the many crises which may have inspired them. These people have been described as 'Ahrensburgian', as the stone tools they employed are very similar to those discovered at an earlier archaeological site from the Ahrensburg valley, northwest of Hamburg, Germany. The defining characteristic of these tools is very specific: flint blades are modified to form tanged projectile points. Certainly, one of the uses of these points was as arrowheads. Excavations by a German archaeologist, Alfred Rust, in 1935–36 unearthed 100 arrows, shafts or fore-shafts from late-glacial lake sediments at the foot of Stellmoor Hill within the Ahrensburg valley. Among these finds, tanged points were found still attached to their wooden shafts. Associated with the finds were a vast number of faunal remains – 17,000 mammal, bird, and fish bones and more than 5,000 reindeer antler remains. Closer examination of the bone remains revealed distinctive damage caused by arrow injuries including fragments of flint embedded in bone. Currently, Stellmoor provides the earliest, direct evidence for hunting with bows and arrows, arguably the greatest revolution in human hunting strategy, a method that remains in use amongst many societies today. Tragically, the arrows recovered from Ahrensburg were destroyed during an air raid over Kiel on 22 May 1944.

Although labelled as Ahrensburgian, there is no direct proven link between these British hunters and those who were hunting in

Germany, but they were certainly employing a very similar technology. We can perhaps visualise hunting parties exploring a slowly warming pre-boreal Britain, carrying bows and arrows, tracking and ambushing reindeer in river valleys and along lake edges traversing a topographical landscape that in many places in Britain remains virtually unaltered to this very day.

Around 11,700 BP, the temperature warmed once again to a cold temperate climate, and Britain became an open, treeless grassland. The effect of the long reign of cold was to retard the regeneration of tree species. Reindeer still thrived, but now the warming climate seems to have benefited wild horses which proliferated in the open grassland environment. The warming climate hastened the melting glaciers elevating the sea level year on year; while we remained the western part of Doggerland, the southwest corner of Britain was starting to develop its distinctive outline as the rising sea inundated the low-lying ground, pushing slowly eastwards and subsuming the channel valley.

People once again followed the game in Britain. Were the people who now followed the wild horses and reindeer descendants of the Federmesser, or the Ahrensburgian explorers who probed Britain's resources in the chill twilight of the Younger Dryas, or were they a different people altogether? This is not clear, although like the Federmesser they seem again to have arrived from the east and, once again, the evidence for their presence in Britain is scant. What little we know of these people comes from a handful of remarkable sites. Once again technology had moved on; tools are recognisable by long flint blades which were very skilfully struck from carefully prepared cores. Most of the sites so far discovered have been found in the southeast and east of Britain. It is a fanciful notion but looking at their distribution I cannot help wondering whether these sites demark two migration routes of reindeer or wild horses that were followed by mobile hunting parties from communities permanently based in Doggerland. If so, one would have led people westward along the south side of the Thames valley, the other northwest through East Anglia. Were they different tribes separated in territory by the Thames, or one tribe, with family groups seasonally exploiting the hunting potential of different territories either side of the Thames, perhaps from a heartland near the confluence of the Rhine, the Thames and other rivers with the Channel River?

The long blades of flint required relatively large, good-quality flint and special stages of preparation. At a site near Herne Bay in Kent, it seems that they persisted in this flint-working method even when the flint was less than optimal, 79 per cent breaking in manufacture, suggesting that these new tools had a very significant purpose or advantage. In many ways the mystery of lithic technologies is that the materials they were shaping with these tools are missing. We have only one piece of the puzzle. Astonishing forensic analysis of the wear patterns on the tools can suggest what materials were being worked and in what manner, but still a gulf of information is missing. The versatile and durable nature of flint and similar materials can create a confusing history of use, particularly as flint tools frequently had many applications. A tool such as a burin, a miniature chisel which has an edge resembling a lathe tool, may show wear from its intended purpose such as scoring a tine of antler, but equally it may have been retouched on its edge to scrape antler wood or skin. Further adding to the confusion, tools used and discarded in one age were sometimes picked up at a much later time and remodelled for yet another purpose. While some long blades clearly show wear patterns associated with butchery, one of the characteristic patterns of wear or damage they exhibit, called bruising, seems to have been commonly experienced in their use. While it has been suggested that this results from the shaping of soft stone hammers, given the multiple purposes these stone tools were put to, it is not impossible that this could have been caused by the action of chopping antler and bone or perhaps even the hardworking of wood.

While the changes in flint tool manufacture are often used to differentiate between the peoples of our past, flint tools in museum cases can seem dull and uninspiring. It is important to remember that they probably represent significant advances in technology, just as in my lifetime hand tools have largely been replaced by portable battery-powered devices, making slow tasks faster and inspiring new ways to fashion materials. By experimentation our ancestors were finding novel ways to fashion tools, to work with new materials and to innovate with familiar materials. Allowing for the strange reality that we sometimes adopt tools as a fad rather than for practical advantage, I believe that it is reasonable to suggest that changes in flint tools during these Palaeolithic explorations of Britain reflect changes in our ecosystem along with advances in our knowledge and use of materials, particularly our most abundant resource: trees.

Along with long blades, these late Palaeolithic hunters left behind small points fashioned from small blades, that have been interpreted as arrowheads. If so, they are evidence of bow-making, a woodworking technique that required a new way of working with wood. But, what role did the long blades play in this story? In 2004, while filming a programme about past lives for the BBC, I teamed up with the late Christopher Boyton, a very dear friend and Britain's finest bowyer. Together we reconstructed a prehistoric longbow using flint tools. Interestingly, of all the tools we experimented with, the one that we continuously gravitated towards was a long blade which enabled the heavy shaving work to be accomplished efficiently, rather like using a blunt draw knife. What I learned from that experience is that should I make another bow with flint tools I would go to great lengths to make a good long blade.

Three Ways Wharf, today lying close to the Grand Union Canal in Uxbridge, West London, between the M40 and Heathrow airport, seems an unlikely site for a Palaeolithic hunting camp. But over 10,000 years ago it was precisely that. Excavations revealed two occasions when hunting parties camped on the bank of the River Colne, which back then would have been a wide impressive wilderness river with gravel bars braiding and flowing into the Thames. Archaeology has revealed that this valley was a popular hunting ground for reindeer and wild horses during the first occasion. Here a small group, perhaps just a few individuals, kindled a fire and cooked two reindeer legs and maintained their equipment.

The Sami in Lapland say that good places to camp are possessed of a special feeling. What fascinates me is that if as time travellers we were to remain camped beside the fire left by that small Long Blade hunting party, late one autumn 500 to 1,000 years later our very long wait would be rewarded by the voices of people once again. They would walk over to the still visible remains of the stone tools and now very aged reindeer bones and feel that this was still a good place to camp. They would pitch a tent and kindle their own fire. Here we would see them bring in red and roe deer that thrived in a landscape that was now significantly more wooded. We would watch as, using flint cobbles exposed by a nearby stream, they fashioned tools to butcher the deer and to process the skins; their flint toolkit is also subtly different. Now we see one of them concentrating, busy shaping a piece of wood by chopping it with a flint adze. As we observe them cook the meat and laugh, two young boys call out as they approach the camp; from the demeanour of the adults, we deduce that they have been away for longer than is considered appropriate. Bows in hand, full of mischief, they know they will be forgiven for bringing back the treasure they are holding aloft: a swan and a duck. From now until today, there will be no more cold interruptions to life in Britain.

These new residents, who today we call Mesolithic, are our last hunter-gatherers: their culture will endure for 5,000 years until supplanted by farming. Standing at the spring of our current age,

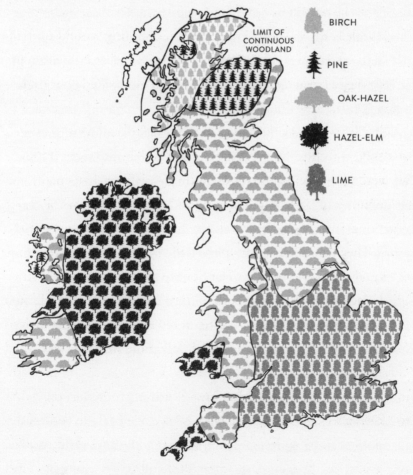

BIRCH

PINE

OAK-HAZEL

HAZEL-ELM

LIME

LIMIT OF CONTINUOUS WOODLAND

Map showing the different kinds of tree
which typically grow across the British Isles

the Holocene, Mesolithic Britons will witness great changes: the dramatic warming of the landscape, the rapid growth of broadleaf forests, advances in wood working and, at around 8,000 BP, dramatic floods and the eventual engulfing of the once mighty Doggerland, isolating us from the rest of the world on Island Britain. We can only imagine that day, when people gazed out at the sea realising that they were now separated from the rest of the world. Fortunately, by then these confident people were tree wealthy, inheritors of a diverse range of native trees (those trees that were growing here before we became an island). If you listen hard, you can almost hear the echo of a Mesolithic voice, saying, 'I know. We'll build a boat.'

Having spent most of my life travelling in wildernesses, many in the boreal environment which in many ways exemplifies post-glacial Britain, I know how difficult life must have been for our ancestors. I sense their dramas, like the horror of a Magdalenian mother discovering a child with frostbitten toes in a world before antibiotics. I know that more recent native societies have made frostbite medicine from the inner bark of larch, but back then the larch was not growing here. I have thought long and hard about the decision making and risk management demonstrated by those early explorers. Understanding the environment and their tool kits, I empathise with their vulnerability, and I am deeply moved by the courage with which they decided to step into the chill unknown. What was it that gave them an edge? What did they look for to suggest that they could manage here?

I believe the answer to those questions is to be found in their knowledge and understanding of trees. Although imperceptible in the early archaeological record, the pollen record reveals there was one tree that was a common denominator in all those early expeditions into frigid Britain. A tree that above all others can make life possible. The birch tree.

This landscape in Northern Finland today may illustrate the Britain our post-glacial explorers knew: an arctic tundra with birch trees, some juniper and the occasional pine tree. It would have been a challenging land to traverse requiring a high level of skill and good clothing.

The sight of a birch wood stirs an ancient memory deep in the heart; the birch tree is our tree of life. Birch trees were the post-glacial pioneers that first clad Britain in forest. They tend to form open woodlands with trees all of the same age, creating nursery conditions for other species such as pine. Relatively short-lived, they rapidly decompose, enriching the ground.

The Pine trees of the Caledonian forest are directly descended from the first pines to grow in Britain after the last glaciation. Once forest like this covered Britain, but with the warming of the Atlantic period the pine forest retreated to the Scottish Highlands. Once covering an estimated 15,000 km², today this uniquely evolved ancient old-growth forest covers just 180 km².

Ancient oak woodland is one of the richest and most complex land environments. Centuries of undisturbed accumulation of decaying leaves and wood have created a rich ecosystem that is home to a multitude of threatened species. These inspiring woodlands are nationally important and yet massively threatened. They need our immediate help for their preservation, maintainance and restoration.

Try to walk through this woodland and you will sink to your knees in the boggy ground. But alder thrives in wet ground, and so such woods are called alder carr. Here the alder has also been coppiced, an ancient system of woodland management that creates trees with multiple stems.

This ancient Beech is a rare sight, a survivor from past centuries, when pigs were pastured in the woodland. The spreading crown results from pollarding, the process by which the topmost branches were trimmed to produce vigorous new growth above the reach of the pastured livestock below.

Untended beech pollards become top heavy and eventually the branches break, allowing the ingress of fungi and bacteria. Prone to such infections, they rapidly die, creating nutrient-enriched glades in the woodland. Here trees of all ages and species can be seen taking advantage of the availability of sunlight. What cannot be seen is the complex mycorrhizal network that underpins the health of the forest and is most developed in ancient woodland.

Woodland offers many hidden benefits. One of the most important is the way that leaf humus and tree roots slow down the flow of water from the land. Prolonging the water's journey to the sea allows for nutrients to be absorbed and creates and refreshes freshwater habitats, benefitting a host of wildlife while serving to reduce soil erosion and downstream flooding.

Alder

The city of Venice sits on top of a hundred or more tiny islands, separated by canals – it is, essentially, a city built in a swamp. Its construction began in the 1300s, and its starting point was timber piles, load-bearing trunks driven six metres into the ground, pushing through the sand to find stability in the underlying clay. More timber was then laid horizontally across the piles, above sea level, to create a base for the forthcoming construction work. The most famous Venetian edifice is probably the Doge's Palace, built in 1340. It is 152 metres long and three storeys high, so its weight must be immense, yet it has been sitting on submerged wood for almost 700 years. Fortunately, there is a wood which is extremely durable in water.

Alder (*Alnus glutinosa*) is a member of the Betulaceae family, the same family as the birch, and is native to Britain and most of Europe, usually found growing in marshes and along rivers, streams and other wet areas. Alder in fact gives its name to a term for wet woodland: alder carr. The tree's lifespan is relatively short at around 60 years, but in death it could outsit eternity: alder wood doesn't rot in water, so long as it is kept wet. It has not just provided Venice with its platform on stilts; in the past, alder has been used for sluice gates, water pumps and water pipes.

Alder thrives in low-nutrient and barren soil where other trees cannot, because the nodules on its roots carry a bacteria – *Frankia alni* – which can 'fix' nitrogen. Fixing nitrogen is the conversion of nitrogen gas into nitrogen compounds like ammonia, and trees use these compounds to create plant tissue and the chlorophyll that is vital to photosynthesis.

The bark is olive-green when young, but cracks and darkens to grey as it ages. The tree's typical height is about 20 metres, occasionally as much as 30 metres, and it usually retains its dark-green, rounded ovate leaves until well into autumn. Every alder tree grows female and male flowers, and they look quite different. The larger but drooping male catkin is up to six centimetres long and is dark yellow-brown. The brighter, smaller female flower is red and begins life as a similar oval to the leaves, but develops into a woody cone, containing seeds; alder is Britain's only deciduous native that grows cones. The alder's fruit sustains several varieties of moths, including the alder kitten, butterflies and crane flies, while bees take its pollen and nectar and goldfinches feed on its reddish-brown seeds. You might still spot the empty cones on the trees come springtime.

In summer, you might also increasingly notice abnormally small and yellowish alder leaves falling early from the trees. These are signs of distress in the alder population. A new hybrid strain of the fungus *Phytophthora* has afflicted alders in Britain in recent years, the infection causing root rot. Other visible signs of alder dieback are brownish spots of 'rust' on the bark, where the tree has been bleeding from stem lesions, along with large numbers of cones and dead branches and twigs.

Alder bark is rich in tannin, the thin growing twigs can be frayed by chewing and used as an expedient toothbrush, and the bark can be collected from saplings and used to tan leather. The straight saplings peeled of their bark and allowed to dry make enduring poles for tepee-shaped shelters. Alder wood is also well suited to friction fire-starting.

Alder Buckthorn

Alder buckthorn (*Frangula alnus*) is one of my favourite trees. A slender shrub-like tree that has today faded from commercial value, it now grows secretly in the shadows of more mature woodland as a discreet, spindly, twisting tree. It is consequently a tree that most people walk past without a second glance, attracting only the attention of savvy tree lovers.

Don't be fooled by the name – alder buckthorn is no relation to alder. It does, however, share alder's affinity for damp soil, flourishing in bogs and on riverbanks, and is a popular food source for birds, bees and butterflies. Also like alder, it's monoecious. Unlike alder, its male and female reproductive parts are found within the same flower, a small, three- to five-millimetre star-shape with five triangular petals of greenish-white.

Alder buckthorn has a dense, tight grained wood with a lemon-yellow colour that carves well. Historically, the tree was thought to safeguard against witchcraft and demons and to protect from poisons and headaches; the colon-stimulating properties of its anthraquinone content also made it useful as a laxative in times long gone. Its most widespread and impactful use, though, was in the manufacture of gunpowder. Indeed, an assessment in 1785 by a Major William Congreve of the charcoal of various woods determined that the three best trees for gunpowder production were alder buckthorn, alder and white willow. Alder buckthorn was especially valued for the evenness of its burn rate, making it ideal for time fuses. Nowadays, of course, the alder buckthorn is more highly prized as a source of barbecue charcoal.

Apple

Malus domestica is, alongside pears, plums and roses, part of the Rosaceae (rose) family. Its wild origins, as *Malus sieversii*, lie in Central Asia, and the apple's colonising progress tracks from ancient Asia Minor to the Middle East, North Africa and southern Europe. The fruit of the tree was so popular that the early Swiss invented wintertime dry-storage for halved apples, while a number of characters in Homer's *Odyssey* (3,100 BP) are noted cultivators. The Romans bred ever-larger, ever-tastier varieties, and they planted pips wherever they conquered. They used dried apple to make relish, or ate the fruit fresh from the tree.

By 6,000 BP, Britain's Neolithic inhabitants were already leaving apple pips for archaeologists to uncover. These were 'wildings', wild crab apples (*Malus sylvestris*), a miniature version of today's, native to the British Isles. The arrival of the Romans brought new apple varieties, many of which struggled in the British climate, including the dessert apple. A thousand years later, following the Norman Conquest, an influx of French culture introduced new types of apples and methods of fermentation. The art and science of cider-milling was quick to take off among Britain's monks, who developed many new varieties, the forebears of modern strains.

When a medieval mix of war, plague and natural disaster hit British apple cultivation and left us relying on imported fruit, Henry VIII looked again to Europe to regrow our skills. Before long, the world's first modern orchard was producing Pippins in Teynham, Kent. As England built its first empire, its colonists exported apples to America on the *Mayflower* and to Australia on the *Bounty*. The Industrial Revolution saw thousands of new strains cultivated, which we then lost again in the drive to standardise what we bought and ate.

We still have crab apples, although not in a form that our ancestors would recognise. Still small, today's varieties are crimson, orange-red or yellow. The fruit grows in small clusters on three to four-metre, green-leaved trees with a long-lasting pink and white blossom, or on trees with copper-coloured leaves and a darker, purplish blossom. Crab apples don't tend to drop to the ground and, when you bite into one, it reveals a fairly sour taste. Cook them and they lose that crabbiness, becoming as sweet as any standard apple.

The typical apple tree is taller than the crab, up to ten metres. It blossoms with bunches of five-petalled white flowers with whispers of pink, a spectacular sight in the late spring and early summer. The edges of its dark-green leaves are serrated, and each oval leaf is fuzzy to the touch, just slightly on top but densely on the underside.

Many birds are as fond of apple trees as humans, helping spread the pips far and wide; you'll often spot a lone apple tree in scrubs and hedgerows or on rough ground and roadsides. They're also a favoured target for sap-sucking insects like mussel scale and aphids and, if you've ever bitten into a maggoty apple, it may have been a victim of codling moths. Their caterpillars burrow through the flesh of the fruit to consume the core. If it wasn't them, it was an apple maggot, a fruit fly that devours the flesh rather than the core.

Ash

Ash (*Fraxinus excelsior*) is a long-lived tree. The average life is a couple of hundred years but, if it's coppiced, it could outlast four centuries. It is part of the olive family (Oleaceae) and its oil is not unlike olive oil. Its straight-grained wood is strong but flexible and exceptionally shock-absorbent, rarely splintering, and you would be hard-pressed to find a wood that can fulfil as many disparate purposes as ash does. I carved my canoe paddles from English ash. They have now taken me thousands of miles through remote wilderness, where they have often been tested by the rigours of travel in the Canadian Shield. While they have picked up some scars, their enduring strength has been a constant support and comfort. Ash is perfect for tools like axes, spades and hammers; weapons like bows, clubs and spears; sports equipment such as hockey sticks and oars; car frames and carriages; bowls, furniture – all have been made, or still are, using timber from the ash tree. Strong bindings can be fashioned from the thinnest shavings. Its stems are good for firewood and charcoal: ashes to ashes. At a push, its green seeds are edible pickled when immature and have an extensive history in herbal medicine, alongside its leaves which make a good compress for an infected wound.

'Old-smelling wood', the Sioux people call it; that and 'weapon wood'. This gives an indication of how widespread ash is. It is Britain's third most common tree and is native throughout Europe but also to Africa and Asia Minor, and is even found in the Arctic Circle. Anywhere that's cool and fertile with plenty of deep soil with good drainage provides a suitable environment.

These are tall trees: 35 metres high when fully grown, some-times 45 metres, with pale brown, silver-marked bark that cracks with age. In summer its leaves can be seen like lace high in the forest canopy. In winter, you'll notice its smooth, flattened twigs and its seed clusters and distinctive velvety, black leaf buds. Come late April or May, the buds will produce the tree's long (as much as 40-centimetre), featherlike leaves. They are pinnately compound: half a dozen pairs of leaflets, light-green ovals on opposite sides of a central stalk, at the end of which is a single leaflet.

Ash leaves will follow the sunlight so each tree's crown may lean towards the sun. Since ash trees tend to grow together, they will often lean in to form a forest roof. It is thanks to this dome that so many wildflowers grow under their auspices. Ash leaves fall early, in late summer or early autumn while they're still green, so the sunlight filters easily through the natural awning to the ground, encouraging dog's mercury, dog violet and wild garlic to flourish around the trees. These wildflowers support hosts of insect life, including the high brown fritillary, Britain's most threatened butterfly.

The tree itself, meanwhile, offers home and nourishment to all kinds of local fauna. Nuthatches, owls, redstarts and woodpeck-ers will all nest in its branches; dormice don't have to weave their own nests when there's hazel thriving beneath the ash; many moss and lichen varieties grow harmlessly on its bark. The cater-pillars of at least 30 moth species find sustenance in the tree's leaves, eight of them being entirely dependent on it, including the centre-barred sallow. Bullfinches eat the seeds, and many deadwood-dependent species of beetle, like the lesser stag, rely on the plentiful supply of decaying wood resulting from the tree's long life. Overall, ash trees support more than 100 different species, and around 30 of these are completely reliant on them.

In the spring, before its leaves appear, purple flowers begin to blossom in spiked clusters on the twig-tips. Ash is dioecious so these might be male or female, but both will not usually be found on the same tree; if they are, they're on different branches. In late summer, the pollinated female flowers mature into fruit, the winged 'keys' that flutter from the trees in winter through to spring and are dispersed on the wind.

About 30 years ago, a fungus from Asia called *Hymenoscyphus fraxineus* was introduced into Europe. Its original hosts, the Manchurian ash and the Chinese ash, had evolved with the fungus, so they had been able to withstand it without too much damage. Our European ashes are less fortunate, and the consequences are already devastating. Britain is projected to lose tens of thousands of ash trees.

The first signs come in summer, as dark spots appear on the leaves, which turn fully black before the afflicted trees shed their now-wilting leaves even sooner than usual. The trees develop dark-brown, diamond-shaped wounds where their branches grow from their trunks, and the bark beneath turns a grey-brown. Epicormic growth is a tree's natural response to damage and stress: inactive buds low on the trunk attempt to grow new shoots. The trees' crowns die back, and the trees wilt and die.

There are tiny hopeful signs: some ash trees in Britain appear to have tolerance to ash dieback, so there's potential for recovery, even if it looks likely to take 50 years.

Aspen

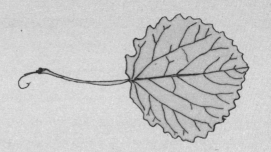

Derived from *aspis*, the Greek word for 'shield', aspen's pale wood is both lightweight and strong. For the ancient Greeks, it conferred more than literal protection in battle: in mythology, the heroic wearer of an aspen crown could venture into Hades and be sure of safe passage back. Christianity's long phase of absorbing or corrupting other religions and traditions condemned aspen as the wood used to build Jesus's cross, although there's just one biblical mention: in some translations, Psalm 137 refers to 'the quaking aspens', giving the tree both its common name, quaking aspen, and its scientific, *Populus tremula*. The leaves tremble and flutter at even a hint of breeze, supposedly a sign of shame for aspen's role in the Crucifixion.

Growing to 25 metres, older aspen trunks can look black, their pitted green-grey bark swathed in lichen. Their slender, dark-brown twigs are bumpily ridged and shiny and have spirals of buds pushed hard up against them. The trees flower in March–April, before coppery leaves open, unusually long (from five to eight centimetres) and with a strangely flattened look to the flexible stalks, as if they've been squashed. The serrated, rounded leaves reach a point at their apex and display very noticeable creamy-white veins. The leaves turn green then fall in a vivid yellow in the autumn.

Aspen leaves and bark can be boiled to produce an antiseptic for treatment of burns and wounds, while the wood's low flammability makes excellent matchwood, reliably extinguishing once discarded.

Aspen wood is light, easily split and odourless – qualities that recommended its use for utensils and tools, particularly dough bowls. Not producing splinters also makes it the favoured choice for sauna benches. Aspen is also well suited to friction fire-starting.

2

GERMINATION

Just three millimetres across, seeds such as this brought
sylvan life to Britain's post-glacial tundra.

t would have been one spring day, sometime around 11,200 BP, when soft, sugary patches of melting snow began to change colour, revealing a russet dusting of tiny seeds that had blown onto the snow the previous autumn.

Each seed was just three to four millimetres across and winged. When viewed up close, they resembled midge-like insects, even complete with two antennae-like protuberances. As is the way with seeds, the majority would perish, their potential unfulfilled. But nature is as persistent as she is miraculous, and it would only take a tiny proportion of those seeds to recolonise the land with trees.

With their protective germination inhibitors washed away by meltwater and the seeds sufficiently hydrated, the cold itself triggered the seeds to germinate. While the multitude of failed seeds washed away in the glacial meltwaters or lay withering and mouldy, slowly recycling on the barren ground, the fortunate minority that settled on soil rooted and started growing. No time could be wasted by these colonisers, as they had only a limited store of energy within their seed. Leaves, nature's amazing solar panels, were deployed immediately to support the fragile seedlings. That summer, a bright emerald haze that would inspire the heart of any artist brought a wash of colour to Britain's tawny, windswept tundra as birch trees reached up to the sun for strength. This was the arboreal dawn of the British woodland we know and love today.

Of the three native species of birch tree, the genus *pendula*, two are particularly well adapted to cold climates: the dwarf birch (*Betula nana*) and the downy birch (*Betula pubescens*). It was probably the first of these, the diminutive and leathery-leaved dwarf birch, that was the vanguard of our post-glacial forests. Dwarf birch is a tenacious tree that thrives in conditions which are impossibly hostile for other species. In sheltered valleys or when the temperature permits, it can grow to a height of just over a metre, but it is most characterised by its ability to grasp the ground and to creep across the land like a wiry heather no taller than ankle height. This is the mountaineer of our tree species. So hardy is this tree that it may well have survived in Britain through the cold spells that had evicted mammalian presence.

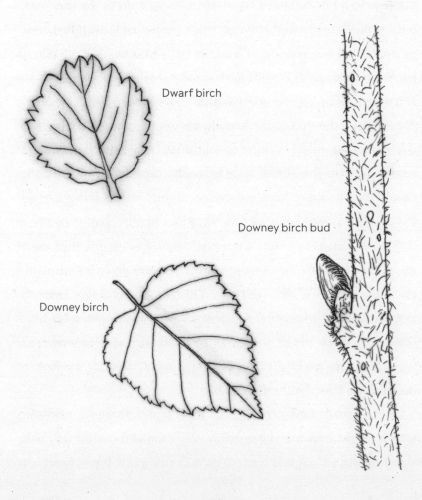

Dwarf birch

Downey birch bud

Downey birch

The diminutive size and multiple stems of dwarf birch technically classify it as a shrub. However, if you have ever travelled on foot across tundra, you quickly come to regard it as a tree, just as the lemming does, valuing the small protection and advantage its groves afford from exposure to the desolate and ceaselessly windy landscape. It is from this perspective that our ancestors would have encountered the birch, when they returned to Britain exploring the tundra emerging from the frosty grip of the ice age. We can only guess at the cultural knowledge and oral memory of those hunting and gathering communities. But when archaeologists visited their long-deserted campsites and investigated them, they learned that they did indeed burn small-diameter firewood of dwarf willow and dwarf birch.

In 2009, I was waiting to load my pack into an SUV outside the Explorer hotel in Yellowknife, the capital of Canada's Northwest Territories. Wisely sheltering from the bone-chilling sleet under the large extended porch of the loading bay was a young Inuit man. He was a hunting guide, waiting to take a party of Americans into the remote bush of the Territories. He looked capable, and it struck me that he had a long-suffering patience about him. As I also stood in the porch, I noticed that he was staring at me; his gaze was intense enough to cause me to feel a little awkward. Eventually, after several minutes of scrutiny, he came over and half-hesitantly stated, 'You're that guy off the TV. The wilderness shows.' I admitted that I was, not sure where the conversation was going. 'I like your shows,' he grinned. 'You tell it as it really is.' 'Thank you,' I replied. Then followed one of those long thoughtful silences that often punctuate conversations in northern culture. Nothing more really needed to be said.

After a while I asked him about his trip and, in short, unadorned sentences, he told me about his hunting expedition. He asked me about my trip, and I told him we were heading north into the barren land, filming. (The barren land is the vast arctic tundra that stretches from the northern edge of the tree line to the high glacial arctic, an incredible wilderness home to the gyrfalcon and the grizzly bear.) As my crew arrived and we started loading up, he said, 'You can burn the live birch out in that country.' He knew that I would appreciate the gift of that knowledge. I thanked him but, before we could talk more, he was called away by the outfitter to carry the bags of the hunters.

I watched him expertly loading the holdalls and heat-shrink sealed slabs of beer cans, one of which was already broken open and realised that he was next to invisible to the excited group of men, each holding a beer, who were crowded around him. Sporting new camouflage hunting clothes, the wrong footwear, and oversized hold-alls, I could hear them talking big, discussing calibres and previous adventures. They would no doubt be both safe and successful because of his Inuit knowledge. As they departed, we exchanged glances and, in that brief encounter I realised that I had met a kindred spirit, a true man of the wilderness. That simple sharing of Inuit wisdom was a potentially lifesaving gift. I wished that we could have spoken longer. I have often wondered about him and I hope that he is well. I have also wondered whether that same advice was once quietly prof-fered by an elder to a party of hunters heading into the tundra that is today Britain.

Dwarf birch does indeed burn green. It needs a small fire of dry kindling to set it ablaze. While there are very few dead branches of dwarf birch to be found, there are usually enough to start the fire. The leaves do not need to be removed, but the density of branches on the fire must be carefully managed to maintain the heat with-out starving the fire of either fuel or air. Dwarf birch burns with a lot of smoke, but the smoke is not unpleasant. It has a sweet, resin-ous aroma that does not taint food but adds flavour to foods cooked directly over the fire. In places where only ankle-high dwarf birch grows, it can sometimes be pulled up to reveal a thicker underground stem that also burns well.

Once ignited, dwarf birch burns enthusiastically, and when fanned by the wind it burns furiously, producing significant heat. What it lacks in size of fuel is made up for by its abundance. The same method was, and indeed still is, used on the arctic fjells of northern Scandinavia by Sami reindeer herders. The dwarf birch has had other uses too. In Alaska, the Yupik people use a tea made from the boiled leaves to treat stomach ailments and intestinal pain (they call dwarf birch *chupuaiya'hak*, meaning 'something you can blow away'), while the Aleut shamans burned dwarf birch to petition for luck when healing sickness. The very young leaves and buds can be eaten raw, and the twigs and buds have been used as herbs to season stews. But if we could ask our ancestors what was the best use of this wiry tree for food, I think they would have explained that it is the best place to snare the rock ptarmigan, which eats the buds where they are exposed and within reach above the snow. Both ptarmigan and grouse remains were among the finds in Gough's Cave in the Cheddar Gorge.

Fire in a frigid land means life. It dispels the lonely isolation one can feel in the vast tundra and makes it possible for humans to explore an otherwise impossibly hostile landscape. As the glacial covering retreated, dwarf birch would have pioneered the growth of trees, enabling exploratory parties of hunter-gatherers to follow the retreating ice closely. In a way, the fruit of those first tiny birch seeds was to facilitate the human repopulating of Britain, a process that has continued without interruption to today.

As the environment continued to warm, the dwarf birch found itself in competition with larger species of birch, forcing it to follow its favoured niche habitat, retreating to the uplands where conditions remained cold and more exposed. Today it still clings to the tops of Britain's mountains, mostly in Scotland, along with its ancient companions, rock ptarmigan and dwarf willow. All of their futures in Britain are threatened as global warming continues to reduce the habitat that they depend upon, literally forcing them ever upwards towards the tops of the highest mountains and eventual oblivion.

I consider myself lucky to have learned to appreciate natural resources from a wilderness perspective. Without that view I would never have made the acquaintance of such a true friend as dwarf birch. Size isn't everything. This is a tree that has sheltered me and kept me warm in parts of the tundra wilderness where no other tree can grow, circumstances which elevate it in stature to equal that of the mighty oak. If you have never seen a dwarf birch, I encourage you to take a long walk in search of one. You will be guaranteed a good view and, as you watch the leathery leaves resisting the unremitting chill breeze, think of our ancestors who repeatedly followed it back onto our soil, at a time when Britain was an unfamiliar, pristine wild landscape, filled with danger and devoid of any easy comfort. I like to reflect on this when I walk down a city high street, surrounded by food vendors, pharmacies, clothes shops and stores selling the latest electronic gadgetry. Our lives are so technologically advanced, yet in many places the landscape itself has only been superficially altered. The hills that once caused our ancestors to sweat as they carried home a load of wild horse meat still cause us to sweat today as we carry home bargains from the shops. That we are a clever species, there is no doubt; look at our machines and our ability to communicate. But we inherited this intelligence from our ancestors, who were equally clever in their age.

Following close on the heels of the dwarf birch came two more robust species of birch: the downy birch (*Betula pubescens*) and the silver birch (*Betula pendula*). They too were wind pollinated and therefore able to reproduce without the need for insects, which were still mostly absent in the frigid post-glacial tundra. The downy birch may well have arrived marginally ahead of the silver birch, as it is more tolerant to cold and can grow in a dwarf-like form in poor conditions. For which reason, it is sometimes mistaken for the dwarf birch with which it can also hybridise.

The downy birch grows to a height of 20 metres. It has a silvery-grey bark and heart-shaped leaves that are toothed along their whole margin and are downy beneath. Its new shoots are smooth and covered with tiny hairs, which makes them characteristically tactile.

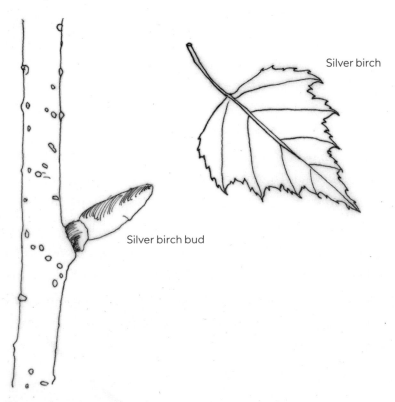

Silver birch

Silver birch bud

By contrast, the silver birch is the largest of our native birches and can grow to 30 metres in height. It has a very white bark with distinctive black markings. Its leaves are narrower, with a more triangular or diamond shape. They lack a toothed edge at the leaf base and have a smooth underside. Silver birch shoots do not have down, but are typically covered in tiny wart-like glands which feel like braille dots.

These species are the birch trees that we are most familiar with today. They are both widely distributed across the United Kingdom, the downy birch favouring the damp margins of streams and marshes, while the silver birch holds sway over the better-drained, warmer locations. It is not uncommon to find them growing alongside each other; the downy birch from a damp ditch and the silver birch from the better-drained ground alongside. In terms of their uses, downy and silver birch trees can mostly be used interchangeably. This is so much the case that, when asked, few people can differentiate between them, or are even aware that they are two separate species.

As befits their pioneering nature, birch trees grow fast. The silver birch is quickest, which may reflect the benefits of its preference for growing in better soil. From pollen records, it has been estimated that the population of birch trees growing as the ice age departed doubled every 50 to 60 years. For our ancestors, this would have been a visibly dramatic transformation, not just to the landscape but to life itself.

As the new forest grew, the whole ecosystem changed. Birch trees provided a food source and habitat for insects, birds and other fauna, even though they are short lived; the average lifespan of a birch tree matches that of humans at around 60–90 years. They also have a remarkable ability to improve soil conditions, reducing the pH and improving soil fertility, which prepares the ground for other tree species and flora to grow.

While birch trees need sunlight for growth, their seedlings do not grow well in the shade of their parent trees, unlike some other tree species. In this way, the birch forest functions to provide nursery protection for species such as pine and oak which, being more shade-tolerant, can grow beneath the dappled shade of the birch canopy to eventually set leaf above it, overshadowing and outstripping the older trees. Likewise, when an oak tree falls, birch trees are among the swiftest species to exploit the light in the resulting glade, so beginning the cycle anew. In this way the birch trees provide a foundation for emerging forest ecosystems.

In the northern hemisphere wherever they grow, birch trees are beyond any doubt humankind's greatest ally, for there is no other tree species that provides us with so much. When I travel in remote northern wilderness, I seek out birch trees, for to be surrounded by them is to be in the company of friends. This is a sentiment I am sure those post-glacial explorers would have understood.

They explored a country that was responding to a rapid increase in temperature. It is believed that open birch woodland predominated for some 2,500 years before giving way to more species-diverse closed woodland. They would have encountered both beautiful monoecious stands of birch and even more biodiverse mixed woodlands rich in other valuable tree species. I truly wish I could have walked alongside them in those pristine forests. The birch tree's ability to regenerate abundantly after felling or fire would also not have escaped their attention. This most versatile tree in every way promotes itself for human use. I wonder, what stories did they tell of the birch tree? Wherever I have encountered first nations in birch country, there are stories about Birch Tree.

The Athabascan people have a legend that tells of a mischievous medicine man, who once decided to test his powers by commanding the wind to blow down the trees around him, species by species. But when it came to Birch Tree, it could not be blown down; it simply bowed over and then sprang back. Frustrated, the mischief-maker commanded the wind to blow more strongly and to snap the trunk of the tree. But, as hard as it could blow, the wind could not succeed. At each attempt, the birch bowed and, when the wind abated, slowly arose to stand tall again. In a rage of temper, the medicine man struck the tree all over to break it. Eventually when he had expended his malice trying to break the tree and was exhausted by the attempt, Birch Tree was still standing. Battered but unsubdued, Birch Tree then spoke to the mischief-maker, sharing the wisdom of its many gifts. Shamed, the medicine man relented, and to this day Birch Tree carries many stripes on its skin that are scars from that event, a reminder to people of the abusive trial that Birch Tree once endured, and features that ever since have marked it out as our greatest tree ally.

When travelling in the far-north forests of the subarctic, it is common to encounter small birch trees that have been bowed over into an arch under the weight of winter snows, but rarely are they broken. The wood of the birch tree perfectly combines the qualities of flexibility and stiffness. More flexible woods such as willow will bow over but lack the stiffness to recover, while less flexible woods simply snap under the strain. These qualities are reinforced by the tree's modest height and its very wide shallow root plate. Its roots are strong, flexible and slightly elastic, giving it a purchase on the soil that anchors it securely, while maintaining the ability to absorb the impact of strong winds. This is a vital quality for this 'standalone' tree that can grow in the open without the sheltering support of other trees.

The birch tree's strong and flexible roots did not escape the inquisitive observations of our ancestors. They found in them an ideal binding material, particularly during the winter when the roots retain their moisture and flexibility while other forms of natural cordage are dry and difficult to access.

Birch roots were harvested from ground that could be easily pulled back to expose the roots, particularly where the soil was sandy or mossy. The roots selected were chosen carefully for their intended use, thin for delicate weaving, medium for lacing baskets, thick for construction. The roots had to be unthreaded from the tangled mat they formed, and then cut cleanly from the main root. Traditionally, only a couple of roots were removed from any one tree, to reduce the impact on the tree. The roots were then trimmed of any unwanted side shoots before being pulled through a brake; a split green wood stick, to strip away the root bark.

Cleaned roots are very beautiful, and much tougher than they may seem. The root can be divided along its length by splitting it from one end and carefully working the split down the root's length. If the split starts to run off to one side, bending back the thicker side more sharply brings it true again. Incidentally, this is a principle that can also be used for splitting shoots of green wood and the roots of other trees. Skilled root workers could skilfully split roots into three using their mouth as a third hand. The roots could be used in the round or split according to need. Usually, they were dried and stored for later use.

Wherever birch trees grow, their roots have been used to stitch together containers and baskets made of birch bark, while thicker roots have been used to bind fish weirs and fish traps. Given that our earliest post-glacial citizens explored a landscape dominated by birch woodland, it would be astonishing if Britain's hunter-gatherers did not also make use of these properties in a similar way.

Birch roots can also be used as a basketry material in their own right. A tradition of birch-root weaving survives in Scandinavia to this day, particularly in Lapland. Thin roots are woven into beautiful coiled baskets, even into complex shapes such as flasks with necks and stoppers to contain salt. In this craft, in addition to drying and resoaking, the roots are also softened and polished by being rubbed between a wooden board and a leather-covered baton. Whole roots form the core of the basket coils and split roots are used for the binding.

These delicately woven birch-root baskets usually have geometric patterns deftly incorporated in the weave. True labours of love, that seamlessly interweave utilitarian function with extreme durability and artistic expression, epitomising the soul of *duodji*, Sami handicraft. Interestingly, these exquisite root baskets are often overlooked by tourists visiting Sami handicraft shops, out-competed by the allure of hand-carved cups of birch wood or Sami knives with decorated antler and birch sheaths and handles.

From the tree's perspective, roots are everything. They provide connection to the soil, anchoring and supporting the tree. Most importantly, they are the means by which it accesses water and the essential nutrients necessary to enable growth and photosynthesis. Although they are hidden from view, roots comprise around one-third of a tree's physical size. The roots closest to the trunk are thick, deep rooting structural roots, that provide the most support to the tree.

Most roots are delicate feeder roots, that fan out from the trunk to collect moisture and nutrients. To maximise efficiency, these roots are shallow and wide-spreading, usually extending two to three times the radius of the trees canopy. The efficiency of the root network is astonishing. Depending on the temperature and humidity, an average birch tree can transport between 250 and 300 litres of water per day. Ninety per cent of that water will be transpired from the leaves, with just ten per cent providing the tree with essential nutrients.

Although hidden from view, the subterranean natural history surrounding tree roots is as fascinating as the natural history within its branches. Perhaps the most mysterious is the mycorrhizal relationship that can form between trees and some fungi. Birch trees are a prime example of this, known to form such relationships with over a hundred different fungal species. A single birch tree may be benefiting from interaction with as many as 30 separate species of fungus at the same moment. Recent research into this intriguing relationship, which is known to have existed for more than 450 million years, is revolutionising modern understanding of trees and forests, challenging many long-held forestry beliefs.

Mycorrhizal relationships begin when fungal spores germinate in the soil, and send out hyphae to a tree's roots. They penetrate the roots and establish a network inside the root for nutrient exchange. They also extend hyphae out into the soil in search of moisture and nutrients that will be fed back to the tree. In this way, the tree's root to soil contact is effectively magnified, sometimes by as much as 700 per cent. But more than simply dramatically increasing the area of soil contact, these hyphae are much smaller in diameter than the host tree's finest roots and root hairs. They can therefore penetrate soil that the host tree's roots cannot, providing access to nutrients that would otherwise be unobtainable.

Some hyphae can even lure, trap and digest microfauna. Through this extensive network, the birch tree is far better able to gather water and nutrients, including phosphorus and nitrogen. The tree transports the water and nutrients through its xylem to its leaves where, through the magic of photosynthesis, they are converted to carbohydrate. Engorged with its mycorrhizally enriched nutrient supply, the tree is able to produce more carbohydrate than it can itself use. The excess is transported back down to the roots and to the fungi, along with some fixed carbon. Thus, the fungi that are unable to manufacture carbohydrate receive their reward. This symbiotic relationship between the tree and the fungus will be a marriage for life.

When these fungi reproduce, they form fruitbodies: the mushroom or toadstool that rises above ground, to disperse their spores. Most notable amongst the mycorrhizal fungi associated with the birch tree is the very easily distinguished fly agaric (*Amanita muscaria*). This is the beautiful fungus with a red cap covered with white spots that is much beloved of children's illustrators and is most often found growing beneath birch trees. It is a toxic member of the *Amanita* genus, which includes both edible and deadly poisonous species, including the death cap (*Amanita phalloides*).

Fortunately, there are also many delicious edible fungi which grow with birch trees. The penny bun (*Boletus edulis*) and the chanterelle (*Cantherellus cibarius*) are the most sought after, while the brown birch bolete (*Leccinum scabrum*) and the orange birch bolete (*Leccinum versipelle*) are also delicious, particularly when preserved by drying.

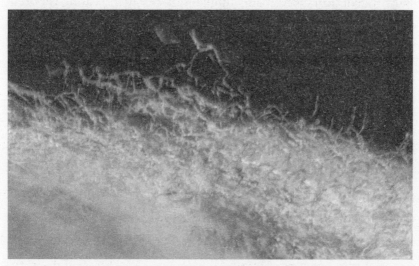

Microscopic mycelium hyphae. Hidden from sight, the fungal mycelium play a vital role in the development of a healthy woodland.

47

Most fascinating of all, scientists have discovered that trees use their mycorrhizal fungi to communicate with each other. They plug into the incredible underground matrix of mycelium, which serves as a means for trees to warn their neighbours of attack by threats, enabling them to increase production of the phytochemicals they use for defence.

There has also been the extraordinary discovery that trees can use the mycorrhizal network to share nutrients between each other, even between species. Veteran trees can share nutrients with their seed-lings and, when the old tree is damaged and in a state of decay, they are able to recycle their nutrients and to share them into the network. Ancient and veteran trees have such long-established mycorrhizal connections that they act as hubs within the woodland network, providing stabilisation within the ecosystem, one reason that such trees are so important to the health of a forest.

Birch trees are early risers from their winter sleep. Their sap begins to rise in late February. By early March it is in full flow, trans-porting nutrients to the developing buds swelling in anticipation of their opening. For a short window, just a few weeks, the sap can be harvested. Birch sap can be drunk fresh as a tonic, or preserved as a lemonade-like drink, or even as a beer or wine.

Every year I make a special pilgrimage to birch woodland to renew my devotion to trees and reaffirm my commitment to protect-ing nature. Although I could tell you that I make this pilgrimage late in February, some years it may be later in March; this year it was the second day in March, for I am intent on intercepting a particu-lar moment in the forest year and nature pays scant regard to the human calendar.

In a more reliable, Aboriginal way, I rely instead on clues in the surrounding woodland. Despite the days being short at this time of year and the nights still cold enough for a frost, the forest is already stirring. After the winter sleep there are always early risers. I look for snowdrops on show, the dog's mercury having set its tiny flowers and primrose flowers just starting to open, with the forest floor green with bluebell leaves harvesting the sunlight. Looking at hazel trees particularly, the catkins are heavy with pollen and the leaf buds will be about the size of a squirrel's toe. It is now, when the first few reliably sunny days warm the land, that I head for the birches. In my pocket I carry a Swiss army knife and a small birchwood cup. On my way I will cut a small growing shoot of elderberry, slightly slimmer than a pencil. I will shave off the green outer bark of this shoot and cut it to the width of my hand. Then using a dead but still strong twig of birch, I will carefully push out the foam-like pith from the elderberry twig to form a perfect wooden tube.

I search now for a mature birch tree growing near to water in a sunny patch of woodland or on a forest margin, looking for a smooth patch of bark about 50 centimetres above the ground. Using the reamer blade on the penknife, I drill into the bark of the tree angling slightly upwards. As I do so, the inner bark will crumble from the hole like potato gratings. As the blade goes deeper, the bark gratings moisten. Now, chamfering one end of the elderberry tube slightly to fit, I push it into the hole and wait. After a few seconds sap will begin to drip from the end of the tube, a tangible flow of life and vitality. It is one of those moments that is really, quite ordinary, but which, after the drought of winter, lifts the soul. Beneath the tube, properly called a spile, I place my cup to catch the drips. While the cup fills, I walk away and enjoy the warmth of the sunlight. This year, above me a great spotted woodpecker was drilling out grubs from a dead ash branch, while high above the canopy a buzzard was also enjoying the sun's warmth, calling with joy as it rode on a thermal into the sky.

Time passes easily at such moments, but not frivolously; contemplation is meaningful. I remember my friends who share a connection to birch trees. Some are wilderness guides, others from northern first nation communities. All will also sip birch sap in the spring.

When I return to my cup it is nearly overflowing, although the flow rate from the spile has more than halved. This reminds me that the sun is now dropping towards the horizon, and of the sway that the sun holds over all life on our planet.

Birch sap at first taste is unremarkable. It tastes like water, with an ever so subtle background sweetness. But it has a freshness that is unique. When you drink it, you feel invigorated. It is a tonic to the body and a stimulant to the spirit. It is hardly surprising for birch sap is full of vitamins that can boost the immune system. Drinking sap unites me to the forest, with the birch tree and with nature. Before leaving, I carve a green birch branch to a tapering point and, removing the elder spile, I plug the hole, sealing it tight. Like all resources in nature, it must not be taken for granted. Over-exploitation of a tree's sap will weaken it and can cause infection. Leaving the tree, the chill of evening is beginning to pierce my clothing, but there is a spring in my step. The birch tree has confirmed for me the return of the growing season. Even if it snows now, it will only be a temporary chill.

Birch sap has been harvested for longer than records exist but, in recent years, the back to nature movement, coupled with easy communication on the internet and social media, has exponentially increased demand, both at a commercial level and by those who want to hike out and harvest their own. It should be remembered that sap is the lifeblood of a tree and that the bark is its skin. Cuts into bark open the tree up to infection by bacteria and particularly by harmful fungi. When we utilise any natural resource, it is incumbent upon us to utilise only what we need.

The most striking feature of a birch tree is its bark. The silver or white outer bark does more than identify it, it proclaims it. Slender white birch trees stand out. They are different from other trees. Their bark shines, peels like an onion and has a grain which runs horizontally rather than vertically.

Birch bark holds many secrets. As a wilderness guide, I have long felt a kinship with the birch tree. This is a tree that lives beyond the edge of forest society and pioneers new ground. A traveller in the wild, it needs to be self-reliant and to go equipped to withstand exposure to the ravages of nature. Just as guides must wear clothing that protects against the rain, sun, wind, cold and insects, so the birch tree has a specialised protective outer layer. The bark of a birch tree is smooth to protect it from harmful parasites, lichens, fungi and moss.

Growing in the open without the shade of a closed forest, birch trees are vulnerable in the midwinter to the effect of overheating by sunlight in the day followed by rapid cooling at night, which can prove fatal. They overcome this by having a thin bark that does not absorb and store heat as thicker barks do but, most significantly, the bark's light colour keeps it cool by reflecting away the sunlight. The silvery white colour of the birch bark results from tiny crystals of betulin, a triterpene that derives its name from the birch tree. It is one of the key chemical components of birch bark. It is so abundant in birch bark that it can comprise up to one-third of its dry weight. It functions as a chemical defence against microbial and fungal infections. It is also hydrophobic and highly flammable.

As birch trees grow in girth, the outer bark layer splits and peels away, often lingering on the bole as a mass of tough, dry, papery curls. These make it unappealing to herbivores and, at the same moment, enable it to slough off harmful mosses, lichen and fungi, while refreshing both its light-reflecting surface and its gas permeability.

In the late winter, the thin, clean bark of the birch allows it to photo-synthesise through the bark itself. This gives the tree a head start in seasonal growth, which, coupled with its excellent gas exchange and transpiration, increases the tree's metabolic rate, facilitating faster growth.

From a human perspective, birch bark is an extraordinary material which can be put to a multitude of uses. Its most well-known use is for fire-lighting; a few curls of naturally peeling birch bark, or bark strips from a fallen dead tree, ignite easily with a match or if scraped to produce a fine fuzz of tiny bark shavings with hot sparks from a ferrocerium rod. Being waterproof, it can be relied upon in bad weather, and it is normal practice for hikers in Scandinavia to keep a few pieces of birch bark in their rucksack or pocket. The birch tree's short lifespan, coupled with its enduring bark, means that there are always fallen birch trees that will supply the fire-starter's needs. It is bad practice to cut bark for fire-lighting from a living tree. Doing so opens the tree to infection and leaves a marked, disrespectful scar. Trees that show signs of such bark-stripping always catch my eye and sadden my heart.

Once our birch trees grew large enough, there is every possibility that the bark was used to make baskets, perhaps even for cooking in prior to the availability of pottery. During his famous journey to the Coppermine River in the late eighteenth century, Samuel Hearne witnessed just such use of birch bark:

… fine level country, in which there was not a hill to be seen, or a stone to be found: so that such of my companions as had not brass kettles, loaded their sledges with stones from some of the last islands, to boil their victuals with in their birch-rind kettles, which will not admit of being exposed to the fire.

They therefore heat stones and drop them into the water in the kettle to make it boil.

Ancient stones, cracked and crazed from heating and called pot boilers by archaeologists, are commonly found in middens. They could have been used in ground ovens, or even inside a sewn up carcass, to cook meat, but equally they may have been used to boil in a vessel of hide or bark.

Worldwide, I have found there are trees described as the tree of life. These are species which are locally abundant, and which offer many uses. The birch tree is our native tree of life. It has more uses than I can even begin to describe here. Genetic analysis suggests that its evolutionary history may stretch back more than 60 million years, a long time to perfect the evolutionary adaptations necessary for its niche in the forest ecosystem. The oldest fossil evidence for a birch tree is a beautifully preserved leaf from a now extinct species, *Betula leopoldae*. It is 49 million years old. As a lover of birch trees, the very sight of that clearly recognisable leaf touches me emotionally in a way that is hard to describe. The birch tree is a symbol of hope and survival.

There was a day when a man skiing across the barren fjells of northern Lapland found that he had overreached himself. Confronted by the discernibly plunging temperature, which was already 40 degrees below freezing, he realised that he must act to find warmth. He felt his vulnerability; a creature evolved to live in the tropics, caught now between the clear sky that exposed him to the dispassionate malevolence of space, that was sucking away his life warmth.

He gambled all to save his life, choosing to follow the frigid air down towards a river in the hope of finding some woodland. Mindful of the danger of rash decisions, he chose a gentle route of descent, for cold is apt to cause clumsiness and clumsiness precipitates accidents; without question, an accident now would have only one outcome. As he descended, the cold gnawed at his mind, as only cold can, numbing reason, poisoning hope with chill despair. But moving was good and the body-generated warmth was a blessing.

He moved efficiently, taking a pause on every ski step as an old Sami had taught him, to avoid generating sweat that would compromise the insulation of his clothing. In his mind he weighed the equipment in his pack – spare clothing, big jacket, kettle and a fatty slab of dried reindeer meat – focusing always on a plan. As he did so, he rounded a spur of the hill. There ahead of him was a small grove, hardly a wood, just weather-tortured saplings; birch trees that had grown their whole lives leaning into the wind.

Picking a fold in the ground where the snow was deepest, he stamped down a trench to get out of the chill breeze. With the *leuku* (a large, lightweight Sami knife) on his belt, he felled a living sapling that was forearm-thick at its base. He stripped off all of the bark and then trimmed the tree of its branches, carefully piling them in bundles 50 centimetres long, graded from thinnest to thickest. Now, thankfully and with practised skill, he constructed a crib of smaller branches and filled it with the birch bark, which he lit with a match.

At this temperature, the bark smoked and crackled but, regardless, it burned. Across these lively flames he carefully laid the wood, starting with the thinnest twigs, gradually building upwards to those which were two fingers thick. He sectioned the trunk and, not risking failure, he split the awkward wiry grained wood before laying that on the fire also. Damp, thick acrid smoke bellowed from the fire and in this he suspended his kettle filled with crystallised icy snow from deep in the trench. Making a seat with his skis, he laid down some birch branches to insulate his feet and pulled on his large coat. In a few minutes he would have bouillon cooking and he knew that, with continued focus, his life was assured. Pausing for a moment and feeling at last some warmth from the fire, he looked up and thanked the birch tree and the knowledge of its uses garnered from earlier generations.

I know this story well, for the experience was mine. It was of course a simple scene played out in the north as it has been played out countless times before in human history. I have no doubt that such an experience was also lived by some of those reindeer hunters exploring a pre-boreal Britain. Perhaps for them, it was, as for me, an experience that taught the true value of the birch tree – our greatest sylvan friend.

Beech

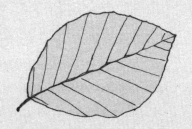

Overhead, a thick leafy canopy frustrates the sunshine, allowing through the smallest shafts of light to reveal the thick layer of withered leaves that blankets the woodland floor. Butterflies dart among the rarest, coralroot bittercress and red helleborine orchids clustering in the deep shade around the trees. For bufftip, ghost, oak-tipped, pale November and peppered moths, this is home, their caterpillars feeding on the tree foliage while birds, mice, squirrels and voles all forage for tree seeds among the fallen leaves and husks. The smooth grey tree bark also hosts truffle fungi, engaged in a simple barter with the trees, swapping their photosynthesised sugar for nutrients, and porcelain fungi and oyster mushrooms hint at unseen decay while bearded tooth fungus grows on the dead fallen trunks and the larger branches, too. The ridged trunks can be enormous; they grow gnarled and knotted with age, opening chances of refuge to hole-nesting birds like great tits, great spotted woodpeckers, jays and siskin. Wood-boring bark beetles find sustenance in the deadwood on and around these long-lived arboreous giants.

The beech tree (*Fagus*) is native to southeast England and southeast Wales wherever the soil is dry and the atmosphere humid. It grows quickly on well-drained soil – limestone and chalk are ideal – and may reach 30, 40, even 45 metres. Despite these heights, beech puts down a somewhat shallow root system that's susceptible to drought. The root plate is easily exposed in strong winds, so storms can bring down beeches in numbers.

Beech timber burns well for fuel, and was used to smoke herring; it still is, among high-end manufacturers hoping to attract those yearning for olden times. Beech soot can be boiled and diluted with water to make a dark-brown pigment, but the timber is probably better used in furniture and equipment. The wood is easily worked, and it's great for rolling pins, wooden spoons and several other kitchen utensils, as well as some tool handles and sports equipment. It's stiffer and stronger than oak and has been used to make drums and piano pin blocks thanks to its impact resistance – it holds a tone somewhere between maple and birch.

The beech tree also works well for hedging because it is marcescent: its leaves wither and die in autumn but large proportions of them do not fall, contributing instead to the huge domed canopy that overshadows the intensely shaded woodland floor amidst groups of beeches. Not many woodland plants can prosper or even survive under a beech canopy.

The long, markedly pointy and red-brown leaf buds resemble small cigars. They develop into lime-green pointed ovals, four to nine centimetres long with five to nine pairs of veins giving a corrugated look. On the edge and underside of each leaf are soft, small hairs, but the leaves darken as they age and lose these silky hairs.

As the leaves open up from the buds, small yellow-green flowers appear, both male catkins dangling on long stalks from twigs and pairs of female flowers, each pair protected by a woody cup. Fruit begins to grow: beech mast, sizable brown nuts in spiky green husks, which eventually split open to release the nuts. We can eat them raw in small quantities, but they're toxic in excess. Roasted beech nuts are a nutritious alternative to coffee.

Bird Cherry

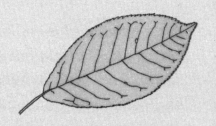

The bark of bird cherry (*Prunus padus*) gives off a nastily acrid pong, which our forebears interpreted as a sign of supernatural powers: they believed that placing some at the front door would keep the plague at bay. It would certainly keep visitors at bay, so it may even have worked.

In Britain, bird cherry trees are a common sight on the banks of rivers and in wet woodlands in Scotland, northern England and Wales, but is rarely found in the south. Its timber is much lighter than wild cherry wood so it is used for making small handles, cabinets and boxes rather than larger furniture. That foul-smelling grey-brown bark is smooth, with no sign of wild cherry's lenticels (horizontal lines of raised pores), and at one time it was commonly used as a pesticide.

Along the twigs lie clusters of buds; their sideways (lateral) growth results in a bushy appearance once the leaves have unfurled. These leaves are serrated, like those of wild cherry but with much sharper teeth. At the top of each leaf stem sit two red glands, secreting nectar that summons ants to ward off any leaf-munching moth caterpillars.

Flowers come out in April, five white petals to each one and several flowers to an initially upright stalk. Those at the base open first; by the time the last opens at the tip, the whole cluster is weighed down and drooping in long trails of white. Like the bark, these flowers emit a very strong scent, but this one is sweet and evocative of almonds.

July brings red-black, bitter-tasting cherries, which ripen by September to be mainly eaten by birds.

Bird cherry was also a traditional choice of wood for carved brushes and shrink boxes.

Blackthorn

Blackthorn (*Prunus spinosa*) is one of my favourite British trees. Growing to seven or eight metres in height, they thrive on moist, well-drained soil, particularly in full sunlight. In neglected fields and pastures they are apt to take over, forming large, dense clumps that gradually join together. When the brilliant white flowers burst like star balls from their buds, I know that spring has arrived. While the weather can of course, still turn cold, these flowers provide a vital life-preserving source of nectar and pollen for our bees.

Blackthorn is aptly named. The branches have a very dark, virtually black bark. They are also covered in long, strong thorns that pose a threat to farmers and foresters managing their growth.

Nasty infections were a common problem for hedge layers working with blackthorn. The leaves are attractive to many caterpillars, providing a ready source of food for the many bird species that take advantage of the protection of the tree's spiny habitat to nest in.

The wood of blackthorn is strong and springy with a stringy grain and a yellow colour, that darkens with age and with a brown heartwood. It was typically used to make long-lasting walking sticks and the famous Irish sheleigly, (the European knobkerrie).

A few flowers added to a salad bring a delicate almond flavour to the dish from the prussic acid they contain. Many tonics and medicinal syrups were historically produced from blackthorn, utilising the bark, flowers and fruit. Today we only use the fruit, mostly to produce sloe gin. However, the fruit can be made edible and can be made into the tastiest wild cordial of any of our native fruits.

Box

The North Downs, a 153-mile chalk-hill range in the southeast of England, runs from Dover in Kent to Farnham in Surrey, and is treasured for two Areas of Outstanding Beauty: the Kent Downs and the Surrey Hills. The latter includes a notable attraction – you could spend several hours climbing Box Hill to its summit. The National Trust has mapped out a 13-kilometre hike to do just that and, when you approach the west-facing slopes, you'll probably notice a sweet scent. That's the smell of the thick, waxy leaves of the wild-growing tree that gives Box Hill its name. They've been established there certainly since the seventeenth century, probably since the thirteenth, and quite possibly for even longer. *Buxus sempervirens*, a member of the Buxaceae family, is a slow-growing but prolific evergreen that can survive for hundreds of years.

Left to itself, box does best in scrub and woodland or on hill-sides, its yellow-green flowers blossoming in April and May. Being among the most popular hedge plants, box topiary is likely to be seen adorning parks and gardens, despite every part of the tree being toxic. Don't swallow anything from a box tree – it will upset your stomach and irritate your skin.

Hard and pale yellow, box timber is Britain's heaviest wood, too heavy to float on water. It was a favoured material among early woodwind-makers because its weight and density lend the instruments more volume and brilliance. It takes four or five centuries to grow a big enough tree, however, and in that time the wood becomes full of imperfections, so boxwood clarinets are today quite rare.

For our ancestors, boxwood's hardness had several advantages: the wood is strong enough to be used as a soft hammer when knapping flint, can be fashioned into sharp, strong projectile points and is an excellent wood for bow-making.

Buckthorn

Also known as purging buckthorn (*Rhamnus cathartica*), the berries of this sturdy tree are a powerful laxative. Western medical practices from the Middle Ages to the eighteenth century were largely preoccupied with ridding the body of illness and disease via bloodletting, vomiting and laxatives so buckthorn was a popular purgative. The berries' properties go beyond laxative: they can also cause constriction of skin cells, perspiration and increased urination, largely because buckthorn berries are slightly toxic to humans and can also cause skin inflammation. These small black berries, like the buckthorn's orange bark, are better used unripe for making a yellow-green ink.

Green when they first emerge from the buckthorn's yellow-green, four-petalled female flowers, the berries first turn red before reaching a ripened bluish-black up to ten millimetres in diameter. These are drupes – soft, fleshy fruits containing a seed inside a hard central stone. The seeds will be dispersed by birds which, in return, will nest amidst the prickly branches of the small, ten-metre trees. The flowers are pollinated by insects, and they nestle among the dark-green glossy leaves which have, in turn, unfurled from conical, scaly, dark-brown leaf buds on long stalks on the thorny twigs. The dense, thorny nature of buckthorn fitted it well to hedging to retain livestock.

An archaic laxative growing on an unusably hard wood – the purging buckthorn may not seem terribly useful beyond its ornamental pleasures. It is, though, the principal food plant for the brimstone butterfly and its caterpillars, which feed on the tree's leaves, while bees collect its nectar and pollen, as do other insects. Perhaps, though, its greatest hidden secret is its superb quality for making friction fire-starting equipment. With an open grain perfect for the purpose, buckthorn was recommended for fire-starting to Roman reconnaissance troops, and is one of my preferred hand drills for fire-starting, particularly when drilled into a hearth of clematis.

3

WAY-FINDING IN THE FOREST

Wood ants proliferate in mature pine woodland. They construct their
dome-shaped nests on the south side of trees. For our ancestors,
they would have been a useful compass to establish direction.

or our Mesolithic ancestors, the changes in our native forests were only slowly felt, each new species of tree quietly adding to the increasing richness of available material resources. But also, imperceptibly, the canopy of our vast pristine forests was becoming thicker, as the deciduous broadleaved trees gradually closed the forest canopy, casting a dark shade on the understorey where only shade-tolerant vegetation could cope. In many places, the once-distant sweeping views of tundra and cold savannah were now gone. In their place a dense forest wilderness stood to challenge our ancestors.

All over the world, people become lost easily in dense woodland. Forest dwellers develop a sharp eye for faint game trails and unique, easily recognisable forest features that can be used as waypoints. But they are no more immune to becoming turned around than you or me, especially when far from known trails. Even today, with the advantage of the Global Positioning System and smartphones, every weekend or bank holiday families become lost in our woodland parks. Our forests are so small now that this poses only an inconvenience of a few hours before they stumble upon a road or building from which they can work out where they are. But imagine the consequences for those ancient Britons traversing a largely wooded wilderness. Everything tends to look the same in a natural forest, particularly where the topography is bland and indistinct, offering no clear feature to steer a course by. Even the most wary woodland navigator can become confused by a sudden change in the forest's appearance, especially when a deluge of rain falls. Today there are very few places in our much-reduced woodlands where such an experience can be felt. But once it was a common pitfall.

I have often wondered how frequently our ancestors were forced to climb a tree on a high point to gaze across the forest canopy in search of the haze of woodsmoke to guide them back to a campsite or community. Having travelled in forest wildernesses around our planet, I have become fascinated by the way the forest peoples stay in contact, from Indigenous people in northern Australia watching the smoke from their neighbours' seasonal burning to the talking drums employed in the central African rainforest. How did our ancient ancestors find their way through the un-surveyed forests that flourished after the last glaciation? While we shall never be certain how we can be sure that they had a method.

Perhaps an answer is to be found in the methods still employed amongst the Indigenous peoples of Australia, who maintain an oral history of their landscape that is central to their religious and cultural beliefs. The Indigenous stories describe in detail the lives and events of ancestral spirits when they roamed the land in a distant past that is often referred to as the Dreamtime. The journeys of these ancestral beings form a network of storylines which criss-cross the Australian continent, even passing offshore. Far from stories for entertainment's sake alone, these narratives represent a survival training for their keepers. They describe in fine detail how to live, from the correct way to harvest and prepare plant and game to eat, to where to find water and other resources. Using these as a guide, the Indigenous man or woman can traverse their country safely and with incredible accuracy. Far more than a mental map connecting distinctive landmarks, these storylines also locate small places of significance in the landscape and most importantly sacred sites where specific religious rites must be observed, where to linger may result in illness. While to modern industrial minds these beliefs may seem primitive, they are not. One has only to witness the Indigenous peoples' ability to find their way across trackless country on a scale that dwarfs Europe, without a paper map, compass, or the support of costly satellites, to appreciate the value and brilliance of such cultural wisdom.

I have several times enjoyed the privilege of being invited to travel along such 'dreaming tracks'. On one memorable occasion, an old lady led me to a place in the central desert where her family had traditionally dug for water. We travelled for two hours into the desert by 4x4, the old lady sat upright in the passenger seat, pointing in a relaxed way to the left or right horizon with the knuckles of her relaxed right hand, signalling that I should keep going by deftly wagging the hand.

Then as we began to parallel a distant escarpment, I could see in her eyes that she was concentrating hard. Eventually she indicated to slow down and then to turn left, away from the escarpment. Eventually she said quietly, 'Here's good.' We left the car and I followed her to a spot in the desert with no rocks or trees anywhere near, nothing to demark it. 'Here,' she said tapping the ground with her metal digging stick. She was very specific: the place indicated was no more than two metres square. Digging with a small entrenching tool, we descended over a metre into the ground and there, as promised, the red desert sand darkened with moisture and, after a little more digging, provided water. As if that was not amazing enough, I later learned that she had never visited the site before. She knew how to find that place solely by reference to the stories she had learned from her mother and aunties. This was a real lesson in the survival value of cultural knowledge.

That old lady was very respectful of tradition and was careful regarding what details she shared with outsiders. Since then, I have learned that many of these journeys are followed using songs associated with specific ancestral beings and their stories, hence the popular term for these trackways, 'songlines'. Walking such a songline, it quickly becomes apparent that it is impossible for a cultural outsider to read and interpret the landscape to the same degree. However, attempting to do so is a powerful means of better understanding this ancient, enduring and beautiful perspective on the world and its inseparable spiritual tie to the land.

The key attributes of the Indigenous route finder are a good memory and a keen eye for the topography. These abilities develop naturally in the normal course of traditional Indigenous life. Interestingly, these same traits are equally important to the skilled navigator employing modern methods of wayfinding. It is a safe assumption that our ancestors also relied upon their memory and observation; perhaps they too once followed long-forgotten storylines across the land that would have connected key identifiable features in the landscape, such as river estuaries, rocky outcrops and distinctive hills, as well as the location of seasonal food resources and perhaps places of historic importance to those long-lost communities. We shall never know if such stories existed or what form they took, but even with such a method to guide a journey, in woodland navigation the need remains for orientation, a means to establish direction to avoid wandering endlessly in circles. In the top end of Australia, distinctively blade-shaped termite mounds can be used to indicate direction, hence their name: magnetic ant hills. But what of Britain?

DIRECTION-FINDING

Of course, the motion of the sun can be used, rising as it does in the east, arcing across the British sky always to the south of us before setting in the west. While the precise bearing of the rising and setting sun changes significantly though the course of a year, at midday when the sun is highest in the sky and at its strongest, the sun is always due south of us. Using a watch, we can easily establish south at midday and even estimate south from other times on the basis that the sun moves 15 degrees along its arc every hour. All we need to do is accurately visualise the arc the sun follows across the horizon on the day. At night we could look for a clearing to locate the North Star, but this is rather impractical as walking in woodland at night is far from convenient and, besides, we may want to be heading home before dark. Moreover, what can we do, day or night, when the skies are cloudy, obscuring sun or stars?

Perhaps you have heard the adage that moss grows on the north side of trees? Well, try to follow that advice and you will quickly find yourself confused. Moss grows where there is sufficient moisture, and this can be on any side of a tree, depending on the local microclimate. Perhaps historically confusion has arisen in differentiating between mosses and lichen. Being slow growing, lichens are more likely to be a reliable indicator of direction than moss. Some lichens certainly do grow more abundantly on the north or south side of trees, depending upon the species. But in practical terms they will not necessarily be found frequently enough and are not consistently reliable enough to traverse truly remote country. Both lichens and mosses respond principally to the local availability of moisture. Moisture levels in any landscape are affected by many factors such as wind currents, water courses, splash zones from waterfalls and permanent bodies of water all coupled with prevailing air currents. Lichens will favour the surface of their substrate that most nearly equates to the optimal for their life support, but this does not guarantee that they will indicate a given cardinal direction. To make true practical use of natural indicators of direction, the indicator chosen must be reliable and consistent to a much higher degree. What is needed is a slow-growing organism that is induced to grow towards a source of influence which is consistent and directionally reliable.

For this we need look no further than the trees themselves. Trees are by their very nature wedded to one of the most reliable influences, sunlight. Every day of their growth, they reach out to sunlight and are themselves shaped over time by the availability of sunlight which is strongest during the midday period, when the sun is at its highest and true south of the tree. Being tall and slow-growing climax species, Britain's three national trees, the English oak, the Scots pine and the sessile oak, are superb natural compasses, as lime trees and elm trees also would have been.

PHOTOTROPISM

A tree's pursuit of light is a miracle. The process begins with the seedling beginning its upward journey in life by detecting and reacting to the force of gravity. Called gravitropism, this causes roots to grow downwards and stems to grow upwards against gravity. Once the shoot can photosynthesise, it responds to the light in a different way, through the process of phototropism, in which light causes the phytohormone auxin to elongate and multiply cells, driving their upwards growth.

Throughout its life, any tree will respond in this way to the availability of light. Light falling strongly on the stem of a tree causes the cells to transport the auxin to the shady side of the stem, where it causes cell elongation. This in turn forces the sunny side of the tree towards the sunlight, also causing the tree to put out new shoots and more leaves on the sunny side. As a tree grows, the lower branches fall into shade and are discarded. If you look up into a tree you may see small dead branches not yet shed. Mostly they will be found on the side of the trunk which received the most light as the tree grew.

Now, this effect can of course be influenced by the presence of shady neighbouring trees or geography. For example, a tree which grows in the shade of a north-facing cliff will only be able to reach out to the north for light. If, however, the tree eventually grows higher than the cliff, it will immediately turn towards the light and reach out more strongly to the south. The great advantage of pine and oak trees is their ability to emerge from the canopy and over-reach the shady influence of other trees and geology. Any tree standing proud of the forest with an unimpeded view of the sky can respond phototropically to the local movement of the sun.

Each day, the sun passes from east to west, the extent of the easterly and westerly horizons varying according to the day of the year as the Earth tilts backwards and forwards on its annual cycle. During the midpart of each day, however, the tree receives its strongest exposure to sunlight. Every day at midday, the sun is due south of the tree. Over the long span of the tree's growth, this causes strong phototropic growth, the cumulative results of which are the longest, heaviest, bushiest and most horizontal branches growing on the sun-facing side of the tree. Branches on the shady side of a tree will tend to be less numerous but, significantly, they will be more vertical and usually attempt to grow around the central stem towards the sunlight. In very old oak trees, it is sometimes found that a horizontal branch has grown towards the sun only to be shaded by a missing and long-decayed neighbour which forced it to grow off course for a time, before once again resuming its true course when beyond the shadow of the obstruction.

Like any such method, it requires practice but, with the investment of time studying trees, this method of direction-keeping is incredibly reliable, particularly when several trees are consulted and attention is paid to the tallest branch tips.

One special circumstance can be found when trees grow in locations where they are exposed to the constant torment of the prevailing wind, an occurrence very common along the coastline. Under these circumstances, the tree may grow bending strongly away from the direction of the prevailing wind. Close consultation of the branch growth will, nevertheless, still reveal more branches reaching horizontally to the sunny south and more vertically to the shady north. If you are aware of the direction from which the prevailing wind blows, such trees can provide two methods of establishing direction.

Where a lone tree has been felled, the phototropic influence can be established from the tree's growth rings, which are wider apart on the shady side of the tree, resulting in the heart of the tree rings being located closest to the sunny side.

A by-product of the increased growth of foliage on the sunny side of a tree is to provide a greater degree of shelter on that side. The ground is frequently drier, although I would not rely upon that as a means of obtaining accurate direction, save under one very specific circumstance: in mature pine woodland, where we have an invertebrate to guide us, the wood ant. At about one centimetre in length, wood ants are easily recognised. They thrive in pine woodland, where they construct large dome-shaped nests of pine needles to establish perfect climatic conditions for their pupae and larvae. For this purpose, they seek out locations which are favourable, nearly always in the shelter on the south side of a tree.

Back in the early 1990s, I was asked to evaluate a new GPS device. These were a novelty back then and much larger than we are used to today. The company representative joined me in an old pine forest to demonstrate its use. While experimenting with the device, which was also equipped with a gyrocompass, I compared its reading to a wood ant nest and was immediately able to determine that it was not giving an accurate reading. The rep was horrified until he realised that he had set the gyro compass to accommodate the magnetic deviation caused by the electromagnetic interference inside his car. He reset the device, and all was good. How much had the device been off course? Seven degrees. That perhaps gives some idea of just how accurate woodland indicators of direction can be.

HOW TO NOT GET LOST

I have lost count of the number of times I have been asked by lost woodland hikers where they are. But I do recall that in nearly every case they had a map and compass with them. This is hardly surprising. The reduced visibility in woodland makes forest navigation one of the most difficult types of direction-finding. Traditionally, the pocket compass is the forester's companion, a simple tool that makes it possible to unerringly establish direction in the shadowy depths of vast woodlands. Those of us who have learned to find our way by dead reckoning through the world's most remote forests, from the immense boreal forests to the towering rainforests, using just a tiny compass needle to indicate the position of north, enjoy a satisfaction that is lost to those who depend solely on battery-operated devices.

After its adoption as a navigational tool sometime in the eleventh or twelfth century, the magnetic compass revolutionised our means of navigation and survey. Certainly, anyone hiking into woodland should carry a good compass and a local topographic map and know how to use them. But there are a few additional considerations that can prevent our becoming totally disoriented:

1. Before setting out into the forest establish the geographical boundaries of the region you will be exploring.

    For example, one of my favourite areas is bounded to the east by a road which runs north–south, to the south by a stream and to the north and west by a railway line which conveniently makes a tight bend in the northwest corner of the woodland. So long as I do not cross a railway line, the stream or a road, I know that I am in the part of the forest I intend to explore. My starting point is a car park on the road close to the river on the southern boundary.

2. Next, rate your boundaries by their safety, by their convenience to use as a handrail back to your starting point and by their ease of discovery.

    In the example above, should I become turned around, I can listen for the sound of traffic or a train to help locate a boundary. Listening for a train, I may even be able to establish where it changes direction. The stream meanders and the ground will be boggy and full of thick, tangled vegetation. From this, I can reason that the rail boundary is easy to find and a reliable indicator of direction, but it is clearly unsafe to use as a handrail. The river is the most awkward ground to walk back to my starting point. The eastern boundary is the safer option and ultimately the best handrail back to my point of origin. But if I head directly east, when I strike the road boundary, I may not be able to determine whether to turn north or south to the car park.

    So, my plan will be to head northeast towards the junction of the rail line and the road. If I strike the rail line before the road, I can simply walk east, parallel to the line, until I strike the road. If I meet the road and not the rail line, the chances are greatly increased that I have struck the road north of my destination. By aiming off from my ultimate destination in this way, I simplify the process. Striking the road boundary, I then need only to turn right and walk south back to my starting point.

As a family, scratch this on the ground as a mud map and discuss the various options. Should one of your party become separated, they have a plan to follow. This leads to positive action reducing fear and panic. Planning in this way immediately demonstrates the value of a compass and of being able to read the trees, and is a wonderful starting point for learning more sophisticated wilderness navigation.

3. Study the map contours to understand the topography. Topo-graphical features change the least over time and are therefore highly reliable for direction-finding. One difficulty in woodland navigation is that small rises and falls in the land that are too small to register on five-metre spaced contour lines seem more significant than they would in more open terrain. Don't be surprised to find such features.

4. Know where you set out from. If it's a car park, remember its name so that you can ask a local for assistance finding it.

5. Know what direction you set out in and try to establish a mental map of the course you have taken.

6. Pay attention to the orientation of the topography you are traversing. By looking through the trees, you can easily read the topographical trends and changes of slope around you. In a forest with a strong slope this may be all that is needed to maintain your orientation.

7. OK, so you were distracted by searching for wild mushrooms and are completely turned around. NO PROBLEM. You know roughly where you are, you know where you need to get back to. Use your compass or the trees to orient yourself and set out to your preferred boundary. To walk in a straight line, look for a feature ahead of you in line with your heading. Walk to it and then take another orientation from the trees or your compass and find another feature on the same heading to walk to. Keep repeating this process until you strike your boundary. If you can find one, pick up a three- or four-metre straight stick to carry in one hand; it greatly helps to keep you on track. Once you locate your boundary, follow it back to your starting point.

If you are feeling bold, go out and practise in some nearby woodland. Take your map compass, and I will even let you have your phone and GPS. Make your plan, then pack away your navigational aids amongst your food and warm clothing and explore the area, wandering as you please. Then, when you are ready find your way back, use the trees to establish the direction to the boundary that you have chosen to lead you back to your starting point. If it doesn't work out, you still have your tech gadgets. I will lay money, though, that you will resent having had to carry them once you have learned to read the trees.

Well, now that we have learned to find our way in the forest, let's go for a wander and discover the uses of our other native trees.

Bullace

It is classed as native to Britain but was actually a Roman import some 2,000 years ago. It is safe to assume the Romans brought it here for home cultivation, as revealed by its Latin-derived scientific name, *Prunus domestica* – literally, 'household plum'.

In the wild, bullace is found in woodlands and hedgerows. It grows from five to ten metres, and has thin, shiny branches and a dark-brown bark. Its wrinkly oval leaves have a slightly fluffy feel to their undersides. These leaves contain naturally occurring toxins which produce hydrogen cyanide when crushed – by, for example, chewing. Like the leaves, the bullace's twigs, bark and stems are not edible.

Come spring, bullace develops five-petalled white flowers, succeeded in late autumn by two varieties of drupes: small, yellow greengages and larger, dark-purple bullace. Bullace ripens in October to November, six weeks later than most fruits. It also grows in prodigious quantities, overloading its thin branches so the hedgerow droops groundward. The oval fruit, a little larger than sloes, is blue, purple or black, so wild plums, blackthorn, damson and bullace share a very similar appearance. All these fruits are prone to interbreeding, making it even trickier to differentiate them. They are all safe to eat, although, again, the seed at the heart of each bullace contains cyanogenic glycosides and should not be eaten. The flesh is safe to eat raw, but has an acidic taste that diminishes only slightly as it fully ripens.

Baked or boiled, though, the fruit loses its sourness. You can use bullace in jelly, chutney, jam, syrup, liqueur or wine, and it even makes a tasty crumble.

Dogwood

Dogwood (*Cornus sanguinea*) is mostly thought of as a hedge-row tree, but it can be prolific, growing in dense clumps on the edge of woodland, particularly where there is slightly damp ground and little disturbance. Growing to ten metres, it has beautiful creamy-white four-petaled flowers that form a delicate cross shape. The leaves, which often grow densely, closely resemble those of buckthorn, however dogwood leaves have a rubbery sap in their veins, which enables the leaf to be carefully separated into two halves attached by the stringy sap. The leaves turn a vivid crimson in the autumn, when dense bunches of non-edible black berries form.

Dogwood derives its name from its old name, dag wood. Dag is an old word for skewer, the manufacture of which the straight smooth twigs and branches were ideally suited. I also favour dogwood shoots for making traditional fishing rods, as they taper slimly and are stiffer than the more commonly employed hazel. Long smooth dogwood shoots also make excellent withes for binding, and straight coppiced shoots are stiff enough to be used for arrow shafts.

In the distant past, a tea was made from dogwood bark to treat pain and fever and was one of the herbal remedies used to treat malaria when it was prevalent in the UK. It contains coronic acid that is a salicylate free analgesic. However, the bark tea is also a laxative. The leaves of dogwood were also used as medicinal poultices to treat wounds. In North America the inner bark from dogwood shoots and the roots was dried and rolled between palms, being very lightly greased with animal fat for mixing with tobacco or other smoking herbs. Such tobacco had narcotic properties and was often associated with ceremonial practices.

Elder

Sometimes known as the 'Judas tree', after the unnamed tree that Judas Iscariot hanged himself from after betraying Jesus, elder (*Sambucus nigra*) was also supposed both to call up and to repel the Devil. The Bible, however, remains entirely silent on the elder tree. The Anglo-Saxon word for 'fire' was *æld*, which is the likely source for the name – dried elder stems make excellent fire drills for friction fire-starting, and pushing the spongy tissue out of an elder stem leaves a hollow tube which was used to blow extra air into a fire to encourage the flames. The Anglo-Saxons also referred to their leaders as *ælderen*, although what this says about hot air and hollow leaders is not recorded.

Elder is another odd-smelling and toxic plant. The white pith inside the green twigs which the Saxons used for kindling is one source of the odour, while the toxicity lies in the seeds, foliage, bark and wood. These are all very poisonous, or fatal if consumed in quantity.

The berries are rich in vitamin C but, uncooked or underripe, will also cause stomach upsets. They won't even taste nice, being disagreeably tart. If they are deep purple, they're ripe; if they're green or red, they're not, and even cooking won't rid them of their toxicity. Both flesh and skin are edible after cooking: just cover them with boiling water, let them soak for an hour, then dry them. They'll expand in the process, so you'll want about two-thirds of the quantity you think you'll need. They can then be paired with blackberries to bake a pie, or combined with pears in a tart or almonds in a cake. Elderberry wine still requires the berries to be immersed in boiling water, but they're then left for several days before kicking off the fermentation with good doses of sugar and yeast. Elderberries ripen between late summer and autumn, so they will usually be ready at the end of September – plenty of time for the wine to be drinkable by Christmas.

The berries are a food source for many mammals, from squirrels to bears, while voles and dormice eat the flowers as well. The flowers need cooking before any consumption and only in low quantities: a supermarket cordial will be no more than four per cent elderflower; flavoured teabags might be ten per cent. Gathering elderflowers, you'll notice a strong scent. You're looking out for cream-coloured petals; if their edges have started to brown, they're past their best. Flowers picked fresh, when the big purple buds have just opened, can be fried in batter to produce fritters. Tempura batter is perfect for this; mix it with sparkling water for extra-light batter and serve warm with honey!

The elder tree's leaves are pinnate, each leaf comprising half a dozen opposing pairs of featherlike oval leaflets which emit another nasty smell if they are touched. The caterpillars of moths, like the buff ermine, the dot, the swallowtail and the white-spotted pug, are all enthusiasts for elder foliage, although something about it repels flies. At one time, elder branches were a common sight hanging in dairies. Folklore held that this kept the milk from turning, presumably without inadvertently summoning Satan.

The elder's short trunk means the trees rarely grow beyond 15 metres high. Its hard, yellowish wood has always been popular for carving and whittling. Its grey-brown bark yields black and brown dye; green and yellow can be extracted from the leaves, and purple and blue from the berries. You can see why elder was once a staple part of the Harris Tweed manufacturing process.

Elms

Sadly, there is only one tale to tell. In 1910, the earliest case of what subsequently became known as Dutch elm disease (DED) was recorded. A fungal disease began killing elms (*Ulmus*) in Europe, their leaves withering months before autumn, then branches dying until, finally, the morbidity reached the trees' root systems. The fungus is spread by elm bark beetles, which burrow into the tree trunk to lay their larvae. This first strain progressed through Europe fairly slowly, not hitting Britain until 1927. It was also relatively mild, taking down around ten per cent of British elms, and the disease had pretty much died out by 1940.

Around that time, however, a new strain from Japan was reported in Europe and North America. It took some time, but during the late 1960s it struck. Markedly more vicious, it arrived in Britain from Canada in 1967, and the country's elm population, like those throughout Europe, was thoroughly devastated: the UK lost 25 million trees. Today, the elm is a pretty rare sight, and usually as a hedgerow shrub rather than a tree.

What did we lose?

WYCH ELM

The wych elm (*Ulmus glabra*) is a true British native. It is much smaller and sturdier than the English elm and established itself as far north as Scotland.

The wych elm has larger leaves than other elms, and its wood is better for carpentry because English elm wood tends to split. Wych elm was a highly favoured bow wood in prehistory, its flexibility and resilience perfectly suiting the purpose. Even in medieval times longbows were fashioned from wych elm to help reduce the excessive demand for yew.

ENGLISH ELM

The English elm (*Ulmus procera*) was probably only native in southern Britain, but increased its range dramatically during the Bronze Age. It was hugely successful throughout Britain, and its dominance of the countryside can frequently be observed in the sketches and paintings by landscape artists like Constable, Landseer and Turner.

Flourishing elms are tall and imposing, and in maturity should reach 30 metres; now, after 50 years of DED, it's rare to see one as tall as 20 metres. Their potential lifespan is a century or more. A young elm has smooth, grey bark which cracks and browns after a couple of decades. Suckers grow at the base of the trunk, in increasing numbers today because they are a tree's survival-instinct response to injury, an attempt to grow more branches and seed more trees. Ironically, given that the entire genus hangs by a thread, these suckers and their promise of new saplings are often a nuisance and need to be controlled.

Elm wood has interlocking grain, making it flexible, tough and resistant to pressure, and also resistant to water. Elm trees thus served unique purposes and were an essential resource for pre-industrial carpenters.

Elm logs were used to fashion water pumps and bored through for water conduits. Medieval elm water pipes are still sometimes unearthed by groundworkers in London and many other British cities. Elm's resistance to water also suited it to many purposes in shipbuilding. Aboard the *Mary Rose* a large elm water pipe was recovered that was part of the ship's emergency bilge pump.

The wood has also been used for floorboards and coffins, and in furniture construction.

Elm seeds are eaten by rabbits, rodents and squirrels. These small mammals have no real trouble finding an elm seed alternative. Similarly, the many moth caterpillars that feed on elm leaves can move on. Not all creatures are so lucky. The tree-dwelling white-letter hairstreak butterfly is wholly reliant on elm foliage for its caterpillars, and its population has been cut by 93 per cent in the last 50 years.

Guelder Rose

Dutch elm disease gained its name because the phytopathologists who first investigated it in 1921 were in Utrecht, not because it had any connection with or origin in the Netherlands. The name 'guelder rose' has a similarly eccentric pedigree, being derived from Gelderland, a Dutch province that might possibly have been where the snowball cultivar originated.

Despite its name's source, the deciduous guelder rose (*Viburnum opulus*) is native to the UK. An upright shrub, it can grow upwards to four metres while also spreading up to five metres. Its coarsely serrated, broad leaves are green when they grow in the spring and red or orange by the time they fall in autumn. Five to ten centimetres long, they have three lobes, and their undersides are finely hairy.

Between May and July, their cream-coloured or occasionally pink flowers open up in clusters, each one surrounded by a circle of larger flowers. Hoverflies love guelder rose flowers, and plenty of other wildlife can be glimpsed among the shrubbery.

In autumn, bunches of vivid red berries emerge, although a yellow version can sometimes be seen on wild growths. When raw, these berries are mildly toxic, but they can be used for jams and jellies after cooking. It could be better to choose something else, however, since guelder rose berries are an essential food for bullfinches and mistle thrushes. Both species are listed as endangered by the RSPB, rated amber and red respectively.

GUELDER ROSE

Our distant ancestors highly valued guelder rose. Its stems have been found woven into fish traps preserved in the silt of the now flooded Doggerland. The stiffness of its straight shoots also made it a popular prehistoric choice for arrow-making.

4

FIRE

A flint-scraper, a marcasite nodule and tinder scraped from a horse's hoof fungus – the classic fire-starting kit used throughout British prehistory.

Whenever I stop at a vantage point in Britain, I wonder how many times our ancestors also looked out across the vista below. Instead of the neat patchwork quilt of agriculture they would have gazed out across the concealing, dark-green forest canopy. Any natural clearings may not have been large enough to spot from a distance. Our ancestors would have scanned the scene, straining their eyes to detect the thin wisp of translucent blue smoke from a distant campfire. The location of such a sighting coupled with their knowledge of the terrain may have enabled them to interpret whose fire it was, where they were and how best to reach them or, if the foraging was poor, how to avoid them. Lords of all they surveyed, they knew how to take care of themselves: they could find their way, make shelter, find food and perhaps even enjoyed the thrill of outwitting bears in risky close encounters. And, most importantly, they could make fire. That is a uniquely human trait and, to my mind, fire is the greatest gift we have received from the forest.

You do not have to be an archaeologist to realise that we have had a long association with fire. One has only to watch an infant staring into flames in fearless fascination. Our bold appreciation of fire is clearly an ancient legacy. The earliest convincing evidence for the hominin use of fire yet found is from the Wonderwerk Cave in the Kuruman Hills, Northern Cape Province, South Africa, where charred bone fragments, ash and heat-fractured stone tools have been found, suggesting that fire was being employed by *Homo erectus* at least a million years ago. This does not confirm that those distant ancestors could make fire, only that they used it, perhaps keeping it burning from natural forest fires.

Astonishing as it may seem, in my own experience I have encountered desert tribes who historically were not able to light fire, but instead always kept a burning ember with them.

The current theory is that fire was probably first harnessed by humans in Africa one and a half to two million years ago, the use spreading northwards. In Europe, the earliest use of fire has been uncovered in Spain at Cueva Negra del Estrecho del Rio Quipar, dated to around 800,000 years ago; hearth sites far enough inside the cave entrance to likely preclude accidental natural fires suggest deliberate fire-starting, though no traces of fire-making apparatus have been found. The earliest evidence for human fires in the British Isles has been found at Beeches Pit in Suffolk, dated to 415,000 BP.

THE NEEDLE IN THE HAYSTACK

While preserved and transported fire will clearly leave no trace at all, the lack of fire-starting equipment in the archaeological record is profound. This is hardly surprising; any such equipment would certainly have been valued and carried away when communities departed from their campsites. If people had kindled their fire by wood friction and left a worn-out firestick behind, the apparatus would have long since rotted away. Even iron pyrite used to produce sparks decomposes, leaving just the flint tools used to strike sparks, strike-a-lights. Yet even here such tools, which can be identified by distinctive wear patterns and very occasionally traces of iron pyrite adhering to them, are rarely found. Anyone who has used this method quickly discovers that any sharp flake or blade of flint can be used for the job so there is no need to shape the flint in any easily recognisable form. Such pieces of flint may also have only been used once, show-ing such little wear damage that they pass undetected. Also, flint is readily repurposed as need dictates. Undoubtedly many strike-a-light tools having been dulled by repeated use and being readily to hand were simply reshaped into scrapers, burins or any other tool of convenience, their tell-tale wear traces being struck off as tiny micro flakes that are difficult to recover and identify.

To find the oldest fire-making equipment we must look to the Middle Palaeolithic in France, where wear-pattern analysis of flint hand axes from multiple locations has convincingly shown that 50,000 years ago the broad, convex faces of bifacial flint hand axes were being used by Neanderthal communities to strike sparks from a mineral, very probably marcasite, a naturally occurring form of iron pyrite. That the pyrite has not been found may again reflect that the key fire-making tool was carried away, or it could simply have corroded. Marcasite is naturally unstable and readily decomposes in a slightly moist atmosphere.

Intriguingly, another mineral that has been found associated with some of these Neanderthal sites is manganese dioxide, rocks of which show extensive deliberate scraping. One theory proposed to explain this activity is that the powdered mineral was being added to tinder to make fire-lighting more certain. Doing so possibly lowered the ignition temperature of the tinder, ideal for use with the low-temperature sparks from pyrite. If that is the case, it only serves to reinforce how vitally important fire was to our ancestors and how bright was their intellect.

Intrigued by the possibility of a prehistoric chemical catalyst for fire-starting, I have conducted a number of my own experiments. The digest of what I have found is that manganese dioxide added to tinder does marginally improve the rate at which the tinder ember takes hold. Adding it as a powder to fungal tinder reduces success by impeding the mechanical contact of the spark with the tinder. I also experimented with friction fire-starting, finding that some manganese dioxide added to the friction dust being produced did certainly assist in the development of a strong hot coal. However, I did not find the value of the catalyst to be of sufficient advantage to make it worth seeking out.

One of the most important finds of fire-lighting equipment was made in the British Isles by Joseph Sinel in 1881, during excavations in the Cotte a la Chévre cave on the island of Jersey. There he found a lump of nodular iron pyrites about the size of a hen's egg, within the remains of a hearth of wood ash and carbonised wood remnants. More information regarding this find is unavailable: the item itself appears to have been lost. But the pyrite was found in association with Neanderthal tool deposits. Many questions are raised; clearly the nodule was brought to the cave and left by people, but where did the nodule originate? Nodular marcasite is most easily found in chalk regions, but Jersey is far from chalk cliffs. Also, why was it abandoned? Was it perhaps deliberately left there in the hearth to be available for use on a future visit?

While the archaeology of fire is faltering, we are lucky that the living example of the Hadza, who have one of the oldest genetic lineages of modern people, is still burning brightly, providing insights and observations of hunter-gatherers living in a way that most of humanity began to abandon 12,000 years ago. I cherish the time I have spent in their company, which has served as the most potent reminder of the humanity of past peoples.

No word that I could hear was spoken, no obvious command, but the men all knew what they were doing. In just moments they had spread out and were vanishing into the savannah woodland ahead of me, like rugby forwards mounting an assault on the touch line. They did not run but stretched out their stride and picked up a fast walking pace. Somewhere ahead there was rapid movement, then there was sudden commotion, baboon alarm calls, dust rising from dashing feet, men calling confidently. As quickly as it began it was over. The small hunting dogs had done their job, surprising a troop of baboons. The hunters had been there to take advantage of the resulting panic, shooting two baboons with poisoned arrows as they tried to flee.

Over in moments, it was a scene that has been endlessly played out in the Rift Valley for millennia. The dead animals were dragged to a convenient shady place and a fire was quickly spun to life with fire sticks. Even before the hunting party had reassembled, it had been decided that one baboon would be taken back to the women and children, the other eaten on the spot. The men suspended their bows and arrows in the branches of the thorn trees. The carcasses were put over the flames and the air filled with acrid, lingering, greasy smoke as the hair was singed off. The successful hunter cut two-centimetre-wide rings of skin from the tail of the baboon and, removing the string from his bow, slid them onto the ends of his bow, visible trophies of his success.

In only a few minutes the meat was sizzling on the embers. In a restaurant, it would have been described as on the blue side of rare. Then it was eaten, the hunters' knives slicing quickly with the speed and jeopardy of a sewing machine needle, expertly dismembering the body that bore too human a resemblance for my comfort. The hunters concentrated on devouring their prey with the same focus they demonstrated in the pursuit, warm globules of fat dripping from their mouths as they feasted. Nothing, absolutely nothing, was wasted; the skull was gnawed clean, then broken open and the brains savoured. As the joints were finished, the bones were cast into the fire and burned to ashes. When the meal was over, all that remained was a bed of hot ashes with two small fragments of scorched bone on top. The baboon now survived only within the bodies of the Hadza hunters who headed off to please their families and to sleep off their gargantuan feast.

As the hunters departed, I lingered and stared at the fire. Although I had witnessed similar scenes many times before, on that day, that campfire for some reason spoke to me. Looking down at the ash, I became acutely aware of how little evidence of the scene that I had just witnessed now remained. In a few days the ash would be scattered by the wind, the bone fragments perhaps moved by animals. The evidence seemed as transient as the wake of a canoe paddle in water. I thought also of those finds from Britain's Palaeolithic past, where our hunter ancestors are known to have sat and repaired their hunting equipment, sites identified by just a few small pieces of flint that remained, impervious to the storms of time. It became clear to me that perhaps 99 per cent of all the fires kindled by our species have simply vanished into dust. Little wonder archaeologists struggle to find definitive evidence for the earliest use of fire by our ancestors, let alone the tools used to kindle them.

The Hadza live in the great baobab forests surrounding Lake Eyasi in northern Tanzania. They have one of the most ancient genetic lineages of modern people and still live to a large degree as they always have, as hunter-gatherers. Year on year, though, this is becoming increasingly difficult, as the trees and plants they depend upon are destroyed by pastoralists encroaching on their traditional homeland. While outsiders often pity their simple existence and seeming poverty, the Hadza do not envy modern society. They are a wonderful people to spend time with, humorous, self-assured, proud of their cultural identity, wealthy in their heritage and their lifestyle. Raising no crops or livestock and preserving no foods, they prefer instead their traditional diet of wild foods, meat, fat, wild honey, many berries and edible roots. For those who have encountered it, the freedom of the Hadza is enviable, the simplicity and directness of their life a survival from an earlier time and a refreshing reminder of what was lost in the move away from hunting and gathering to agriculture. Living as they do, in a landscape shared with dangerous wildlife, they depend upon fire in a way that the developed world has largely forgotten, but which was an experience we all once shared.

When he walks in his forest, a Hadza man walks with immense confidence. He carries with him his essential tools for life, his survival kit. At his waist is his knife, in his hands his bow and a handful of specialised arrows. Among the arrows there is also a fire drill, but rarely if ever the hearth. That he can make as need dictates from one of many suitable dead branches to be found in the forest.

In daylight, he walks safe from dangerous wildlife through his alertness, his intimate knowledge of his country and the security his weaponry provides. But after the sun has fallen, when human eyesight pales in comparison to that of the lion, the leopard and other predators of the night, it is fire that is his saviour. A Hadza man once explained to me that at night I should find a high cliff outcrop, put a wall of rock to my back and keep a bright campfire burning between myself and the dark of the night. Wise advice I have several times heeded. Until you have slept out under the stars in the African bush when ears wring every detail from the sounds of the night and the imagination must be sent on a sabbatical, it is difficult to appreciate the astonishing support that a fire provides. Even if travelling with a big game rifle and a torch, at night it is a simple campfire that remains our best defence. If you must spend the night in the African bush, it is also best done in company so that a fire sentry can be kept throughout the night. Alone, you quickly come to realise the great advantage of a canine companion. With a dog, our ancestors could sleep beside their fire certain that they would be alerted to the close approach of a predator to the dwindling fire in sufficient time to rekindle a blaze of flames.

The oldest remains of a fully modern human found in northwest Europe is a small section of maxilla, upper jawbone, with three teeth still intact. Discovered in 1927 during excavations at Kent's Cavern, Torquay, in Devon, it has been dated to between 44,200 and 41,500 years old. At that time, our distant ancestors shared their world with Neanderthal people. They would also have shared many attitudes and experiences familiar to the Hadza, for while the Britain they knew was cold, quite unlike equatorial Africa, it was also a savannah landscape, across which roamed many dangerous large animals: cave bears that could stand upright to three metres height, dwarfing today's European brown bears; cave lions that were similar to African lions of today, but again were larger; scimitar-toothed cats, woolly mammoths, woolly rhinoceros and hyenas also much larger than their African cousins today. That our ancestors lives were filled with risk is further reinforced by that fragmentary jawbone. Its surface shows evidence of pitting from stomach acid, most likely from the stomach of a hyena. I wonder, did the owner of those teeth stray from the firelight one night? Do not ever let anyone tell you that time travel is impossible. The moment we leave the light of our campfire and step into the shadows, we lose our technological edge and step back a million years to the vulnerability of our pre-fire past.

But fire has protected humanity from far more than just dangerous animals. For the first explorers to explore Britain as the land emerged fitfully from the grip of the last ice age, fire-making would have been an essential skill for survival. Having taught arctic survival skills for 35 years, I have a profound understanding of the difficulties those Late Palaeolithic Britons would have encountered. Looking at their simple toolkit, I am deeply in awe of them. They may not have had our technological advantages, but they were clever, resourceful people and, I suspect, very enduring. Britain at that time would have resembled the landscape that is found today above the treeline on the fjells in the north of Lapland. Then, just as now, there were three essential components for survival in such a landscape: effective clothing to preserve the body's warmth, collaborative effort and the ability to unfailingly make fire at will and swiftly.

So how did our ancestors ignite their fires in such a landscape?

No direct evidence of Upper Palaeolithic fire-starting has yet been found in Britain. But in Europe considerable fire-starting evidence has been found from sites, directly related to the tool-based cultures that are believed to have explored Britain at the end of the ice age – the Magdelanian and Federmesser. The finds all support the use of sparks containing both flint-scrapers and some marcasite nodules. Given that fire-lighters mostly keep their equipment with them, I believe it is safe to infer that this method was also being used in Britain at that time. Plenty of evidence for this method has been found from later periods, the Mesolithic and particularly the Neolithic, during which period men were buried with strike-a-light fire-making kits, perhaps to warm them on their journey into the afterlife. With the adoption of cremation in the Bronze Age, these finds disappear.

THE STRIKE-A-LIGHT

To make a strike-a-light, our ancestors needed three components: iron pyrite, a flint-scraper and a suitable tinder to catch from the spark. It can be fun to make a set today.

Iron pyrite is found as marcasite on the coast, where it erodes from sea cliffs, particularly chalk cliffs. When found, it looks like a rusty lump of iron, either with a surface that is either irregularly crystal like or smooth and polished. While sparks can sometimes be struck from the surface of such a nodule, it is usually best to break the nodule into two halves by striking it on a large rock or hammering it between two rocks. Care must be exercised here to prevent injuries from fragments that can break away with great force. Broken open, the best nodules will show a shiny metallic interior, with long metallic crystals radiating out from the centre of the nodule. It will require no further preparation.

The flint-scraper can be any blade or flake of flint with sharp surfaces. Ideally it needs to be thick and robust to withstand its use.

The tinder is the most difficult component to obtain. During the Upper Palaeolithic, this would have been a critical material to carry and would have needed to be preserved from moisture. We do not know what tinder was being used in the Upper Palaeolithic but, following the example of Inuit communities who were still using this method within the last 200 years, we can envisage the employment of a mixture combining very finely powdered charcoal mixed into one or all of the following, dried and very finely shredded moss, dwarf willow down, and bog cotton down.

By the Mesolithic, more tinder alternatives existed, as the Star Carr archaeological site in North Yorkshire proved. At Star Carr, both marcasite nodules and flint strike-a-lights were found, but perhaps more significantly the remains of a variety of fungi suited to use as tinder were also found: horse's hoof fungus (*Fomes fomentarius*), willow bracket fungus (*Phellinus igniarius*), cramp ball fungus (*Daldinia concentrica*) and birch polypore fungus (*Fomitopsis betulina*). All these fungi can be used as tinder, although the birch polypore is poor in this regard, and I imagine was gathered for its other properties most likely its medicinal benefits.

The horse's hoof fungus is a renowned tinder that was used for fire-starting in prehistoric Britain.

The willow bracket is an often overlooked tinder fungus which grows mostly on willow, although sometimes on alder and birch. Like the horse's hoof fungus it has a deep trama layer and was prized for fire-starting. Many other bracket fungi can also be used to provide tinder for a strike-a-light.

Found growing on dead and decaying ash trees, cramp ball
fungi can be ignited with sparks and glow like a charcoal briquet.
Broken open, the distinctive concentric rings are visible.

The cramp ball must be gathered dry. It is found growing on
dead ash trees. The perfect place to search for it is on the underside of
a fallen ash tree where it thrives and is protected from rainfall. It need
only be broken open to reveal its internal concentric silvery rings
which can be lightly scraped into a more powdery surface to be used.
A good spark falling in these will catch quickly, igniting the whole
fungus into an intensely hot coal. Both the horse's hoof and willow
bracket fungi are renowned as tinder fungi. It is the spongey, felt-like
trama layer just beneath the cuticle at the apex of the fruit body that
provides the tinder. This can be used intact, although the best results
come from scraping the trama layer to produce a small pile of cotton-
wool like fibres.

Significantly, some of the horse's hoof fungi recovered at Star Carr
had been deliberately processed to remove the top half, presumably to

furnish tinder. Intriguingly, some also showed signs of burning, quite possibly indicating their use to transport fire or perhaps to provide a handy source of heat to melt a thermoplastic resin adhesive, rather like using a heat gun on hot-melt glue.

Given the clear availability of such a range of superb tinder fungi, it is not surprising that seed downs suitable for use as tinder, such as rosebay willow herb, thistle down, greater reedmace down or clematis down, were absent. From my perspective, however, a particularly brilliant tinder fungus that occurs in birch woodland was missing,

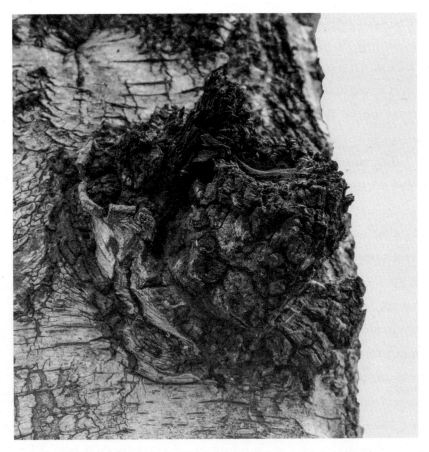

Growing on birch, the birch canker fungus is an unlikely looking fire-starter. However, the interior is a superlative tinder often called touch wood.

the birch canker (*Inonotus obliquus*), today best known by its Russian name, *chaga*. This fungus erupts from beneath the bark of a birch tree like a crusty black tumour. Cut out from the host tree the inner part is woody and bright orange in colour, it needs no special preparation for use, readily accepting a good spark on its orange surface.

In use, any of these tinders will catch from a spark, the tiny first glow of an ember can then be inspired by gently whispering a breath onto it. Once it has enlarged to an ember the size of a marble, it is added to a mass of finely shredded dry grass, dry bark or dead bracken leaves and blown to a flame. Juniper, one of our early native trees, provides a particularly good bark that peels naturally from the surface of the trunk. I have found that this can even be ignited when slightly damp and once the fire has been ignited can be extinguished to be saved and used several times over.

There are several methods of producing sparks from the marcasite that can be used. The most common is to strike the exposed surface of the marcasite with a light downwards glancing blow with the flint-scraper. This will drive off small, hot, bright red sparks. Alternatively, sparks can be produced by scraping the marcasite quickly with the scraper, using considerable pressure. Personally, I prefer to strike the marcasite. I find the process quick and reliable, constantly providing fresh sharp facets on the flint-scraper, although my own method is different yet again. I hold the flint-scraper still and strike it a glancing downwards blow with the marcasite. I find that this causes the sparks to shower downwards in the most predictable way making it easy to aim them into the tinder. Marcasite easily produces sparks daily. Left idle it oxidises and becomes less effective.

But imagine the drama of the past, when certainly at some point our ancestors found themselves caught out in the open in the teeth of an oncoming blizzard, cold hands fumbling to scrape up a small pile of gossamer light tinder from a piece of well-protected horse's hoof trama and then, despite a lifetime's practice, came the struggle to cast dull red sparks into the tinder, when life itself was in the balance. How the heart must have risen when a warm, slightly sweet musty aroma was suddenly smelled, a sure sign that a spark had caught in the tinder and was growing.

FRICTION FIRE

It is certainly the case that fire-making using sparks was historically very prevalent in the colder, damper regions of the world. But this does not preclude the possibility that our ancestors could also make fire by harnessing the friction heat from rubbing wood against wood. Friction fire-lighting is a widely practised way of producing fire that spans the globe and remains in widespread use today. That it is widely absent from the archaeological record reflects the ephemeral nature of the apparatus. When it does show up in archaeology, it can be surprising. Even Tutankhamun took his bow drill friction fire-starting set with him into the hereafter. Not a shiny new bow and drill fire starting set, but a rather well-worn set.

Was this his own set? Did that wood work particularly well? Questions perhaps only a veteran friction fire-starter would ask. I have been making fire by friction for nearly five decades. I know where my current firesticks are, but I have no recollection where the many hundreds if not thousands I have made in my lifetime have gone. They have vanished. What I can say with certainty, though, is that friction fire-starting, where suitable woods can be obtained, is as reliable as any other method, is more widely available and is easy to learn.

There are many ways in which friction can be produced with wood to make fire, but the absolute best way to make fire by friction in the cold and damp is to drill one piece of wood, the *drill*, into another piece, the *hearth*. Drilling with a suitable piece of wood with sufficient speed and pressure generates considerable heat and a resulting hot dust. When sufficient dust collects, it coalesces and begins to combust, forming a self-sustaining glowing ember that can in the usual way be placed within a bundle of dry fibrous tinder and blown or wafted to flame.

In the cold and damp, the most efficient way to achieve this is to employ a relatively thick drill, thumb thickness. This produces enough friction heat to overcome moisture and the ambient cold. Using a thick drill precludes spinning the drill between the palms of the hands, requiring it instead to be rotated by the use of a driving belt or thong that can be held with toggles at each end for two-person operation or fastened within a bent piece of wood or antler for solo use. Downwards pressure is achieved using a hard bearing block of stone, bone or wood, which can be held in the mouth or hand.

This method of fire-starting was not just used in ancient Egypt; it also had wide use in the High Arctic by the Inuit, who, living in a treeless land, utilised driftwood for the purpose. Driftwood works especially well for friction fire-starting, as some chemical transformation seems to result from the wood's exposure to sea water. It is totally conceivable that our ancestors were employing just such a method to start their fires during the Upper Palaeolithic if not much earlier. Moreover, if I am faced with the need to make fire from nature, I can easily do so by friction but will have to travel to a source of marcasite if I am to make fire by sparks. To my mind, the only advantage in fire-starting with marcasite and flint is the relative compactness of the kit, which easily fits into a pocket-sized pouch.

As our forests flourished during the Holocene, so too the available woods for fire-lighting improved. While theoretically any wood can be used for friction fire-starting, in practice some prove to be too hard, requiring an effort that even an Olympic athlete would struggle to maintain. What is needed is a wood that quickly and easily produces hot friction dust. Ideal choices in Britain's forest are:

Native	Non-native
Lime	Horse chestnut
Ivy	Sycamore
White willow	
Alder	
Juniper	
Pussy willow	
Hazel	

This method is relatively simple and will be familiar to members of the Scout movement worldwide. Selecting from one of the above wood choices for the drill and hearth, look for a dry, dead, still-standing branch that will furnish both components. The wood should be not yet quite at the point of structural weakening. The hearth is first carved into a flat board no less in width than twice the diameter of the drill, 60 millimetres, and the same thickness as the drill diameter, a thumb thick 28–30 millimetres.

The drill is carved perfectly straight and cylindrical ø 28–30 millimetres and sharpened at its drilling tip to a cone-shaped point with an angle of 120°. At the non-drilling end, it is tapered for the last six centimetres to half the drill diameter and a similarly angled point made. The length of the drill should be a full handspan in length, around 24 centimetres. The bearing block can be carved from green wood, 10 centimetres long with a ø 50 millimetres. Split this in half longitudinally and carve a depression in the middle of the flat surface to receive the top of the drill. The driving thong was historically fashioned from a rawhide leather strap; today, modern cordage is used. But it can easily be made from multiple strips of willow bark loosely twisted together.

At one end of the hearth, carve a shallow depression to receive the drill tip and then, anchoring the board to the ground under your non-dominant foot, twist the drill into the thong, seat the drilling tip into the hearth, support the top of the drill with the bearing block held firmly against your shinbone and rotate the drill with the bow. It is a little awkward at first attempt, but with perseverance it becomes easier.

Drill to seat the drill tip into the hearth, continuing until the drill has burned a circle into the hearth the full diameter of the drill. Next carve a wedge out of the edge of the board reaching to the centre of the seated depression, such that one-eighth of the depression is removed.

Now prepare your tinder. Place a piece of bark or a wood shaving beneath the carved notch to cradle the friction dust and begin drilling. Drill with sufficient speed and pressure to generate smoke and look for dark brown fine wood powder beginning to fill the notch. Maintain the speed and pressure until the notch is filled with dust, which should be smoking independently of your drilling. Stop drilling and carefully roll the hearth away from the ember, being careful not to break it. Waft the ember with your hand until it begins to glow. At this point, carefully transfer it to your dry tinder and blow it to flame. Easy! Well, it will require some patience and practice.

HAND DRILL

The hand drill, as its name suggests, is simply spun between outstretched palms. The easiest way to make fire with a hand drill is to use a straight dead shoot of hazel or elderberry and to drill rapidly into a lump of birch canker. Doing so swiftly ignites the fungus into a glowing ember; it is an easy process that takes only a few seconds. So long as you have the fungus, the drill can mostly be found as needed. This method would have worked in a cold climate but archaeologically would leave absolutely no trace.

As you will already have guessed, there is absolutely no evidence that the hand drill was ever employed in Britain but, as the Mesolithic advanced, the temperature warmed and the possibility to use a hand drill for fire-lighting increased. This is the simplest drilling method and to my mind the most elegant of all friction fire-starting techniques. The hand drill is conveniently portable: being no larger than an arrow, it has a long history of being carried by hunters in their quiver of arrows. In some parts of the world, an arrow even served as a hand drill; in others, the bow as a hearth. Certainly, Roman reconnaissance troops were trained in its use as a survival skill and may have employed it in Britain.

The most usual hand drill technique is similar to the bow and drill method, drilling into a hearth board with a cut edge notch to collect hot friction dust. It is an ancient method of fire-starting. Mostly the drill is cut green, straightened and dried to be carried for repeated use. The hearth can similarly be cut green and dried or can be made from dry deadwood on demand.

When I was a young boy, I was fascinated by ancient fire-making. I read everything I could find on the subject. One anthropology textbook I encountered suggested that our ancestors may have discovered wood friction fire-starting by observing trees rubbing together in nature. Strangely a few years later I saw a clematis vine rubbing against a wild cherry. It was smoking and scorched. I have never seen anything like it since and, had I not witnessed it with my own eyes, I would never have believed it possible. Naturally, I immediately experimented with both woods. Cherry was unimpressive, but clematis was brilliant and remains one of my favourite woods for fire drills. I felt that I had discovered it, but many years later I would discover that others had of course beaten me to the discovery, even the ancient Greeks 2,500 years before.

There is no practical reason today to use a hand drill, but to my mind little beats carrying and using one to light my campfire. Doing so connects me directly to my beloved trees and fungi, to my ancestors, to the ancient past and to my many friends in indigenous communities both past and present who also have twirled sticks for fire. It is an honest way to make fire, a fair and balanced exchange of energy.

But I am not alone in this attitude, a Ju/'hoansi bushman I once worked with revealed to me that he carried three means of making fires in his steenbok-skin hunting bag. He had matches to light his tobacco pipe that was fashioned from an old 50-calibre cartridge; he had a flint and steel set made from an old file which he carried with a tinder of fungus in a container cunningly repurposed from a dead D cell battery; and he had a set of fire sticks made from manketti wood in his quiver with his poison arrows. These were a cultural item, fundamental to his identity as a bushman and essential for lighting fires of spiritual importance, for example for trance-dance healing ceremonies.

If you would like to try out a hand drill, from the list of drill woods, seek out a straight shoot of little finger diameter and 70–80 centimetres in length. Peel off the bark then carefully shave away any side shoots or protuberances so that the wood is smooth. As the wood dries, gently straighten out any kinks by hand.

Meanwhile, from the list of hearth woods, select a dead and straight branch about 40 centimetres long and as thick as the drill. In width it can be just wider than the drill or wide enough to drill notches progressively down both sides. Flatten it into a board. Once dried, the drill can be seated into the hearth in the same manner as the bow drill above. However, now you have only the motive power provided by your palms to spin the drill. It sometimes helps to add a pinch of sand or fine grit to the depression to get the drill tip to begin scorching the hearth.

Sit on the ground with one of your legs straight out and the other folded, as though you are sitting cross-legged. Secure the hearth to the ground with the outer edge of your folded foot, with a shaving under the notch to lift the ember you are about to create. Spit onto your palms and rub them to create good friction, then, placing the drill into the depression, begin to spin it between your palms exerting a firm downwards pressure.

A hand drill is used by twirling the drill rapidly between outstretched palms while maintaining sufficient constant downward pressure. If such an apparatus was used in our prehistory, it is unlikely to have survived.

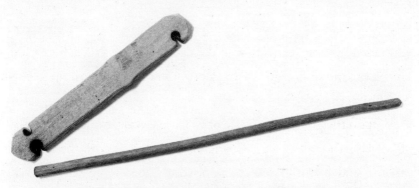

A hand drill fire-making kit. It works well in Britain, but no evidence survives of any such equipment in our archaeology.

Your palms will pass down the drill when they near the hearth, while keeping the hot drill tip constantly seated in the hearth, move your hands back up to the top of the drill and repeat the process. Do not press your palms together too hard or you will develop blisters. Concentrate more on generating speed than pressure. With practice, you will learn to quickly create smoke and powder. Aim to crescendo, starting gently at first then building up your speed and pressure as the drill heats up. As before, you are looking to see the notch fill with dust and for a curl of smoke from its base that indicates that the ember has coalesced and become self-sustaining. At this point stop drilling, gather your tinder and carefully transfer the ember to the tinder and blow it to create a flame.

Drill woods	Hearth woods
Elderberry	Alder
Dogrose	Ivy
Mullein	Clematis
Buddleia	Buckthorn
Buckthorn	Sycamore
White willow	White willow

A sight our ancestors would have known well. A glowing coal produced by friction or sparks is converted to flame by placing it in finely buffed dry fibres, here clematis bark, and then blowing until the tinder ignites.

When you succeed with any of these methods, smile. You have earned it. You have just joined an ancient society and broken your dependency on modern industry to supply you with fire, our most important tool.

MAKING A FIRE

To convert your flame into a fire you need to have already prepared your fire to receive the flame. To do this, choose a place where you have permission to light a fire, then carefully follow the following six steps.

Step One:

Choose flat ground with mineral soil and no living vegetation, away from trees and combustible scrub. Clear away the dead leaves to create a space for the fire. Make the diameter three times the diameter of the planned campfire.

Step Two:

Construct a small platform of deadwood, to provide a heart to the fire, improve ventilation and protect the kindling from the damp, cold ground.

Step Three:

Next, collect a large bundle of tiny dead match-thick twigs that you will find suspended in the branches of trees. These are called snapwood, but do not snap these short; leave them as long as your forearm. The bundle should be thick enough that you can only just encompass it with both hands. Divide this bundle into two halves and cross them over the platform.

Step Four:

Beside the fire platform, pile up other dead, dry branches for fuel sorted by size: pencil-thick, finger-thick, progressing up to wrist-thick. Next, ignite your tinder and introduce it to the space created beneath the pile, where the snapwood bundles cross. If necessary, pull the sticks over the bundle.

Step Five:

Do not interfere with the fire. Allow the tinder bundle to ignite the snapwood. All being well, smoke will rise from the wood, accelerating in speed.

Step Six:

Once flames emerge from the small twigs, cross them with a good handful of the pencil-deadwood, and finger-thick deadwood crossing that. Continue building the fire in this way until it is burning with the size of firewood that you desire.

Once established, let the fire subside to a manageable star-shaped arrangement. The fire can now be adjusted at will to any purpose, kept small to boil and maintain a hot kettle, or enlarged to a bed of hot embers to grill meat. Managing a fire is an art that is a pure joy to learn.

When it comes to extinguishing the fire, try always to plan ahead, so that all the fuel has been consumed to ash prior to extinguishing the flames. Spread out the embers so that they have a flat profile. This immediately suppresses and retards a fire. As the remaining embers cool, collect water and use it to dowse the fire. Thoroughly soak the ashes and embers, stirring them with a stick. Use plenty of water. Any remaining burning sticks can be plunged into a billycan of water to extinguish them.

Once you are confident it is out, use your bare hands to collect up the dead embers and ashes. If the fire is out, this is no problem. If not, you will immediately know that it needs more water. Only in this way can you be sure that the fire is properly extinguished. The cold, wet embers and ashes can now be widely scattered into the under-growth. This having been done, dowse the ground as well, piercing it ten centimetres deep with a sharpened stick to allow the water to penetrate. Now brush back the forest detritus removed before the fire was kindled and restore the scene to the way it looked before you began. It is tough on archaeologists, but kind to the forest.

For humanity, the discovery of fire was a revolution. In addition to dispelling dangerous animals and keeping us warm, it cooked our food, rendering it safer and more digestible. It has been argued that this improved our nutrition and powered the development of the human brain. Fire also allowed us to modify materials, softening resins, hardening wood, improving the working quality of some flintlike rocks.

We used it to make medicinal ashes and to detoxify the poisons in some plant foods and to improve their flavour. Firelight also length-

The Eagle Oak in the Knightwood Inclosure of the New Forest. It seems mysteriously clad in yew leaves, which belong to a yew tree that grows alongside the venerable oak. Here in the winter of 1810, just prior to their becoming extinct, one of Britain's last white-tailed eagles was shot by a keeper. Happily, however, the species was recently reintroduced to the Isle of White, and the tree has survived to witness its silhouette once again pass overhead.

At 400m above sea level Wistman's Wood is one of Britain's highest oak woodlands. The ancient trees, growing in a contorted dwarf form to only 4.5m in height, are believed to be a remnant of prehistoric woodland that once covered all of Dartmoor. The shape of the branches coupled with the moist environment favour the growth of epiphytic mosses, lichens and polypody. Like an oasis in the moorland expanse, woodlands such as this are important habitats for migrant bird species such as pied flycatchers and cuckoos.

The force of life is strong within the forest. Even when a growing tree falls or is blown down, its life story is not necessarily over. If the roots still function the tree can re-root and continue growing, producing a multitude of fresh shoots. The joy of ancient woodland for me is their example of vitality, tenacity and enduring spirit.

It is only during the depth of winter that the importance of a fire truly becomes apparent. The ability to make fire quickly and without fail would have been an essential life skill for travellers in the wild forests of early post-glacial Britain. Our ancestors were inheritors of fire-starting abilities that had been supercharged by coping with the Ice Age.

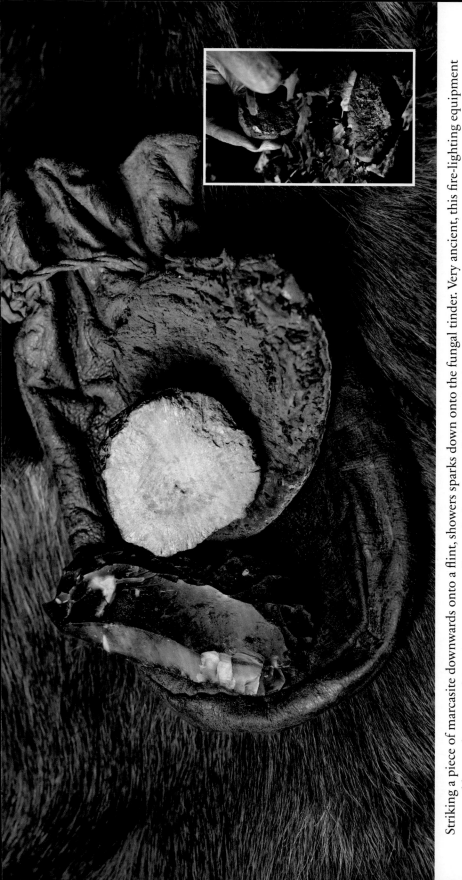

Striking a piece of marcasite downwards onto a flint, showers sparks down onto the fungal tinder. Very ancient, this fire-lighting equipment was used by our hunter-gatherer ancestors as well as early farmers until the availability of iron.

Horses hoof fungus and chaga are superb sources of tinder, and both occur on birch trees. In the frigid birch woodland of the late upper Palaeolithic, they would have been a critical resource for survival. While no chaga has been discovered in archaeological sites, at Star Carr, marcasite, flint strike-a-lights and the discarded lower halves from missing info bracket fungi have been found. The missing portions of the fungi are where the tinder is found.

Although no evidence survives from British prehistory to support the use of friction fire starting, it is highly likely that such methods were known and used. Even the hand drill shown here works well in Britain.

Astonishingly well-preserved yew bows from the wreck of the Tudor warship *The Mary Rose*. The horn nocks that once protected the tips from the string have been lost, but otherwise they look as though they were made yesterday. On some bows even the maker's marks can be seen. Their recovery revolutionised our understanding of medieval archery; some bows such as this one, made from a stave with idiosyncratic grain, demonstrate the great skill of the Tudor bowyers.

ened the human day enabling us to continue working long after sundown, stimulating debate, ideas, humour, storytelling and cultural and religious beliefs. Smoke from campfires enabled communities of mobile hunter-gatherers to locate each other in a vast empty wilderness, and it was around the fire that we conducted our earliest chemical experiments, discovering metals and how to work them. From there, it is but a short hop to the moon landings and the internet.

But fire has one truly profound quality that is so often overlooked. A campfire is the brightest outpost of humanity in the wilderness. It has the unique ability to draw people together and to encourage them to respect and assist each other. I am sure that many of the world's political problems would be resolved if leaders sat around a campfire – a place of equality where each in turn is tested by the smoke and will feel the need to contribute in the group effort to find firewood.

But while our story of fire is one of great human ingenuity and achievement, I like to remember that fire is only possible because of trees. Fire is a gift from the forest. The greatest gift we have ever been given. One that demands a spirit of partnership with our arboreal cousins – a partnership we do not honour equally. It is my heartfelt hope that in rediscovering our early relationship with fire, we can see beyond the blinding dazzle of our technology and re-establish a more respectful and harmonious relationship with the forest, our planet and with nature herself.

Hawthorn

Traditionally a May Day staple, hawthorn is associated with both fertility and death in folklore. When our medieval predecessors fretted that taking hawthorn into their homes was trouble because it 'smelled of plague', they were perhaps on to more than they knew. Hawthorn blossom in fact contains trimethylamine, a chemical that quickly forms when animal tissue starts to decompose, redolent of bad breath or rotting fish.

Its May Day association, adorning Maypoles and woven into garlands, doubtless derives from the fact that hawthorn blossoms and fruits in May, as spring turns into summer. In addition, its leaves unfurl with the beginning of spring, before other plants' greenery really takes off, making it a reliable indicator of the changing seasons. It's a species that flourishes best in full sunshine so the arrival of longer, brighter days is just what it needs.

Common hawthorn (*Crataegus monogyna*) is also happy in most soils, growing up to 15 metres high, and in the wild is generally found in hedgerows, scrub and woodland. It has a knotty and cracked grey-brown bark and slender, brown, thorn-covered twigs. Hawthorn's yellowish-brown wood is tough and very hard, so it is widely used to make boat parts, boxes, cabinets and handles – small-scale carpentry only, since the tree itself is fairly small. As firewood, it burns hot producing a very fine powdery white ash, and it's a good source of charcoal.

Its lobed leaves are pale green, turning to yellow in autumn, and a thriving hawthorn will be densely packed with them. It will attract lots of bird species to nest amidst the thick foliage. When the blossom appears in May, the five-petalled flowers are white or, sometimes, pink. They grow in plentiful clusters, adding to the density of the foliage. As ever, plenty of insects, over 300 species of them, rush in to pollinate the flowers, and deep-red berries called haws soon emerge among the thorns and petals.

Redwings, thrushes and some small mammals eat the raw haws. We cook haws to make wine or jelly. Still-ripening hawthorn buds and leaves can also be eaten and go well in a green salad.

A long list of moth species' caterpillars have the hawthorn as their primary food source, not least the hawthorn moth, along with fruitlet-mining tortrix, lackey, lappet, light emerald, orchard ermine, pear leaf blister, rhomboid tortrix, small eggar and vapourer moths.

Few of these creatures are quite exclusive to the common hawthorn, being equally happy to feast on the Midland hawthorn, and perhaps unable to tell them apart. The two hawthorn species are all but identical, and they frequently cross-breed to produce new hawthorn hybrids.

Hazel

In Chapter 5 we will cover wild-growing hazelnuts, and curse the grey squirrel for being willing to snatch them before they are ripe. Something else to note about hazelnuts is that the UK no longer grows its own, aside from their cultivated cousins, Kent-grown cobnuts. Large-scale production of hazelnuts faded away over 100 years ago so our hazelnuts are now mainly imports. Perhaps that makes it yet more worthwhile to seek them out in the wild.

British native common hazel (*Corylus avellana*) grows in the understoreys of ash, birch and oak woodland and in hedgerows and scrub. Its lifespan is 80 years, but coppicing can prolong its life to as much as a couple of hundred years. Every five to ten years, the hazel tree is harshly pruned almost to ground level in late spring, and that yields a supply of long, straight stems. Hazel timber is exceptionally bendy at that time of year, to the extent that it doesn't break when it's twisted or tied in a knot. These supple stems can be used for thatching spars, in fence and trellis construction and as plant and hedge supports, along with making baskets, broom handles, walking sticks and much more. The hazel, meanwhile, is forced to grow a new stem or stems, which stunts its upwards growth to around three metres but extends its width as it develops into a multi-stemmed tree. Coppicing hazel has a significant role in managing woodland and creates an ideal habitat for many butterfly species, notably fritillaries. Coppiced hazel also offers shelter to nightingales, nightjars, yellowhammers and other ground-nesting birds.

Left to its own devices, a hazel's trunk, probably covered in lichen, liverwort and moss, will grow to 12 metres in height. In the soil around the tree grow milk-cap mushrooms. Like wild cherry, hazel's grey-brown bark is shiny and has lines of horizontal lenticels encircling it.

Green buds start to appear in November, flattened on one side. Hazel is another hermaphrodite, each tree growing both male and female flowers, which unfurl from December to April. These flowers provide an early pollen source for bees, although they struggle with it and can only take small amounts because it's not sticky. Quite the reverse, in fact: individual grains of pollen repel each other, which assists their dispersal in the wind.

The male flowers are on yellow catkins which, from mid-February, hang from the branches in clusters. One catkin may have more than 200 male flowers. These catkins will fall from the tree once they have released their pollen. Waiting in flower buds, six flowers to a bud, on the branches above the catkins, the tiny red females are waiting for the wind to bring the pollen to them.

By mid-May, the hazel tree will be displaying green, cordate (heart-shaped) leaves. Each leaf has a sharply pointed leaf, but the underside is coated with fine white hairs so it feels soft. Barred umber, large emerald, nut-tree tussock and small white wave moth caterpillars all feast on hazel leaves. The leaves that survive the caterpillars turn yellow before their autumn fall.

Come July, each flower bud has grown into a cluster of nuts in woody shells, protected by leafy cups called involucres or husks. The nuts ripen during the summer and into September, when some start to drop from the tree. Many will stay on the tree all winter, as the husks, still bearing the nuts, change to a dark-brown colour.

From late summer, the race is on. Squirrels have a head start, simply because they're unconcerned about ripeness. They are followed by jays, nuthatches, wood pigeons and woodpeckers, as well as dormice, aka hazel dormice. They fatten themselves on hazelnuts ahead of their hibernation; in spring, they happily snack on the caterpillars they find among the hazel leaves.

In addition to common hazel, at least 14 further species of hazel plus several hybrids are known worldwide. One is also found in this country: the larger Turkish hazel (*Corylus colurna*). Native to Iran, southeast Europe and Turkey, it was introduced to Britain in 1582, brought here with a view to commercial cultivation of its edible nuts. In Britain's cooler climate, though, it did not crop well, producing hard and small nuts that did not sell.

Reaching heights up to 20 metres, it is much sturdier and taller than the bushier *avellana*. Its buds, flowers and nuts all look very similar to those of common hazel, but it has a long, straight trunk with cracked bark, and its involucres have long prickles. Its leaves are a similar shape to common hazel's, but the undersides have fewer hairs and the surfaces are glossy, while the edges are more lobed.

Lack of commercial value left Turkish hazel as a chiefly decorative tree, which can often be seen lining city streets. It is tolerant of air pollution, so it thrives in these urban environs.

Hazel can be used for friction fire-starting.

Holly

In winter, you might notice a holly (*Ilex aquifolium*) plant brimming with bright red berries when every neighbouring tree has clearly been stripped of its fruit. If you stop for a moment and take stock, the odds are you'll spot a largish bird sitting atop a nearby tree with a good view of the abundant holly. It's on guard duty.

The mistle thrush is a famously aggressive defender of its food resources. Every October, mistle thrushes select the trees that will become their larders, competing to drive one another away and establish ownership. The bird's attention then turns to keeping intruders away, and it defends by attacking. Tail flicking, wings thrashing, it deploys two sound modes: loud rattling and menacing silence. It's also selective when it comes to choosing who to chase off: great tits, greenfinches and starlings don't feed off holly trees so they're left in peace; blackbirds, finches and waxwings do. It's big for a thrush so chasing them away presents little challenge, but a mistle thrush will even chase off a woodpigeon. Eventually, all these birds get the message, and the mistle thrush takes to its vantage point.

And then it leaves the berries on the tree and feeds mainly on worms. Holly can maintain its fruit throughout winter, long-lasting berries staying on the tree for as long as nine months. When the mistle thrush starts nesting in March, it has already built up a substantial food source for its offspring.

That food source will get to know a lot of mistle thrushes in its lifetime – evergreen holly trees can survive for as long as 300 years. Up to 15 metres tall, they have smooth, silvery bark with small, brown, wartlike blisters, and their stems are dark brown. Everybody is familiar with their dark and glossy leaves, but perhaps not everyone realises that only younger plants have those famously spiky leaves – older trees tend to display smoother leaves. White, four-petalled holly flowers, which can bloom throughout the spring and the beginning of summer, turn into those vivid scarlet berries. The timber is a very white, fine-grained wood, and it's heavy and hard. Stained and polished, it is often used to make walking sticks, as well as in furniture-making and engraving. As firewood, it gives off a very strong heat.

Holly branches have been used to decorate homes in winter for millennia. The inhabitants of northern Europe have always held a party in December, a week-long chance in the depths of winter to give thanks for making it through another year and to celebrate surviving the latest cold snap. The Romans had *Saturnalia*, beginning on 17 December, in honour of their god Saturn, whose sacred plant was holly. It retained its glossy dark-green leaves while others around it were shedding, which was seen as a symbol of fertility, birth and regrowth. There were fairly few locally available evergreens for household decoration, so most other religions and cultures in the region sported near-identical beliefs about holly. When Emperor Constantine converted Rome and its empire to Christianity, holly's absorption into Christian tradition was easily effected. A supportive tale was later concocted, during which Mary, Joseph and the infant Christ hid from Herod's soldiers behind a holly bush.

Hornbeam

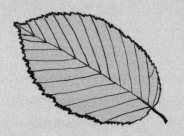

The hardest wood of any tree in Europe is the pallid creamy-white timber of the hornbeam (*Carpinus betulus*). It was the traditional timber used to make yokes to shackle together groups of ploughing oxen, a wooden beam being fastened to their horns to keep them on track. The Romans used hornbeam to build chariots, which needed to be extremely tough, given that they would be both raced and deployed in military skirmishes.

Through the centuries, hornbeam has been relied on in the construction of watermills and windmills and for piano hammers, coach wheels and butchers' chopping blocks, and it is still used for flooring and furniture-making. Pollarded or coppiced, it supplies sturdy and hardy poles. It is also a great firewood.

Hornbeam is our hottest firewood, its seasoned logs burning slowly with an intense heat. For this reason, it was highly valued in the production of charcoal to supply the early iron industry. Hornbeam woodlands grow prolifically throughout the weald of Sussex and Kent, the former heartland of iron production. However, left on the forest floor, despite its hardness, hornbeam quickly succumbs to rot fungi.

There is apparently a tradition in Valenciennes, Hauts-de-France, of leaving a branch of hornbeam at the front door of your beloved, presumably to demonstrate the strength of your ardour, or maybe just to demonstrate your strength. A magical hornbeam potion could supposedly relieve you from drowsiness or even exhaustion.

Worldwide, there are 30 to 40 different species of hornbeam, most of which are to be found in East Asia, while just two of them are seen in Europe. Our native variety, we call common hornbeam (*Carpinus betulus*), and it grows naturally in oak woodland. It reaches full maturity when it is 50 years old, as tall as 30 metres, and with perhaps another 250 years of life ahead of it.

Being monoecious, a hornbeam tree hosts both male and wind-pollinated female catkins. Pollination transforms the latter into samaras, a papery fruit with green wings, clusters of which hang from the branches throughout the autumn. Each samara houses a seed, a nut no larger than six millimetres, cushioned by a three-lobed leafy bract.

The tree's grey bark is vertically marked and, as it ages, its short trunk becomes ridged. Hornbeam's slightly shorter version of the beech tree's cigar-like buds have curved tips and grow pressed tightly to brown-grey, hairy twigs. Its leaves, too, are reminiscent of those of the beech, having a similar oval shape and pointed tips, but they are a little smaller and have a suggestion of pleats and doubly serrated rather than wavy edges. In fact, a hornbeam's leaves offer the shortest route to confirming that it is not a beech. It is classed as a deciduous tree and, although its leaves become a golden-yellow in the autumn, most of them stay on all year.

Hornbeam leaves were once used to treat wounds, and they are capable of stopping bleeding. But the hornbeam's year-round leaves are most effective at providing shelter and sustenance for the local fauna. Unsurprisingly, hornbeam is also the foodplant for the caterpillars of several moth species. Blackbirds, chaffinches, thrushes and wrens will all roost and nest in the foliage and eat the seeds in autumn. The ground around the tree can provide excellent foraging for bank voles, common shrews, field voles, pygmy shrews, wood mice and many other small mammals.

Horse Chestnut

Horse chestnut seems such a quintessentially British tree, yet it is actually an immigrant from the Balkan Peninsula. It is native to Greece and Albania and was first brought from Turkey to Britain in the sixteenth century. Its scientific name is *Aesculus hippocastanum,* which sadly does not mean 'Aeschylus' nuts': while *hippocastanum* means chestnut, *Aesculus* is derived from the Latin for an edible acorn.

What is quintessentially British is the game of conkers. No other country's children collect a tree's fallen fruit with the sole intention of making a hole through its middle, threading it with string then taking it into battle. The game is not entirely unknown beyond these shores, but only here has it been a playground tradition for generations. It was first recorded on the Isle of Wight in 1848, but a similar game was played using hazelnuts in the fifteenth century, so conkers probably evolved from that.

The horse chestnut tree seldom grows in woodland but is routinely found in the parks and streets of cities, towns and villages. Every town has a Chestnut Avenue or a Chestnut Drive, and the chances are that a September walk down that particular road will involve crunching and squelching through countless empty green husks while dodging the occasional missile dropping from a branch high overhead. These husks have spikes and have been protecting the red-brown seed since August. Often, these seedcases open on the tree and the hard seed drops; just as frequently, the husk splits as it hits the ground. Once the conker is released, it might be foraged by squirrels. Most other mammals, including humans, will ignore it since it's poisonous to them. Only deer and wild boar have any great appetite for a conker, and it's rare to see either on our streets.

The large leaves of the horse chestnut are just as instantly recognisable as its conkers. Each leaf comprises five, six or seven pointed and serrated leaflets extending from a central stem. This type of leaf resembles the palm of a hand, hence it is known as a compound palmate, compound because it has multiple leaflets. This is food for the caterpillars of a couple of moth species, the triangle moth and the horse chestnut leaf miner. The latter tend to swarm through the tree in great numbers, which can make the leaves brown and fall early and give the tree a somewhat sorry-for-itself look, but it is good news for blue tits, which feed on the caterpillars.

Before the conkers start to develop in June, April and May bring the stunning blossoming of horse chestnut's white flowers: four or five white petals with a hint of pink at their base and fringed edges. The flowers are, of course, a valuable nectar and pollen source for insects, and they are pollinated by bees. Fertilised, the flowers lose their petals and begin to turn into seeds and cases.

While the pale wood is too soft for most purposes, it is very good for carving. Otherwise, it is once again the conker that has been found useful. Conkers have been used as an ingredient in detergents and shampoos. A few Victorian recipes unwisely suggested making presumably bitter-tasting ground-conker flour.

Dead horse chestnut wood makes a superior friction fire-starting set, and the leaves can be crushed with warm (not hot) water to make a soap, although care must be exercised in case of an allergy.

5

FOOD

Acorns were a staple food of the past. They are rich in tannin
that had to be washed out of them before consumption.

love to walk in the forest throughout the year, observing the quality of the light illuminating the trunks, listening for the tell-tale sounds that betray wild creatures hiding in the canopy, scenting the complex odours of raw nature. I find such observations deeply stimulating. Certainly, a woodland walk can never be boring. As I pass through the dappled shade of the trees, I will occasionally pause to remember the faces and voices of the many tribal communities I have visited.

What I did not realise before I set out on those journeys was just how profoundly they would shape my own outlook, altering forever my perception of my homeland. Particularly my Indigenous teachers in northern Australia taught me to look beyond the four seasons by which we define our year and to see instead many smaller significant seasons within the natural calendar. Each is defined by the availability of a natural resource and can be predicted by the heralding of another synchronous natural event, both united by shared environmental conditions rather than a date in the calendar. For example, when the emerging leaves of hazel trees are about the size of a grey squirrel's toe (4 x 2.5 mm), I know that the sap is rising in the birch trees. Or, when hawthorn trees flower, that pignut stems are well above ground and, with their spindly leaves spreading, are easy to spot. Actually, each of these signal events connects to several significant seasonal events. I am sure that our own ancestors looked on the natural world in the same way, for there is a common denominator that links my Indigenous friends' calendar to my calendar and to that of our ancestors. It is one simple question that hunter-gatherers must ask daily: 'What can I eat here today?'

On the face of it, finding food in the wild is a simple process but, as any honest student of survival soon discovers, the reality of the process is wide of the theory. Sustaining oneself with only wild foods is a very difficult task which requires extraordinary skill, knowledge and experience. I have been asking the question for over forty years, and the answers I have so far found have made me look far more closely at the natural world, to wonder at the astonishing bounty of nature and to bow in respect to our distant ancestors.

Foraging is becoming an increasingly popular pastime, but today we only forage for culinary interest; there is no requirement to subsist from nature as our ancestors once did. But let us consider for a moment what would be required if we were to attempt to do so. What did it take for our ancestors to obtain their 'daily bread'? Begin by listing all of the things you have eaten in the past week. For most of us it is an astonishing quantity. Now imagine our hunting, gathering ancestors' world, one in which they had no potato, rice, pasta, bread, corn, oats, out-of-season fruit and vegetables, spices, sugar beet, chocolate, milk, cheese or other dairy products ... and the list continues. Suddenly the scale of their daily task becomes apparent. Then imagine the added demand of feeding hungry children.

To survive – let alone thrive, as they did – required our ancestors to have an adaptable attitude to their food supply totally unfamiliar to us today. Perhaps at times the forest would have been filled with the sound of foraging occurring on an almost industrial scale. From the scant evidence that survives, such as coastal midden mounds (waste heaps), it is clear they made extensive use of easily gathered littoral foods, from tiny periwinkles to beach-stranded whales, while their stone tool assemblages bear testimony to extensive hunting, revealing a wide use of various projectile and considerable processing of skins into leather.

But what of their forest foods? Here, once again, organic decay denies us most of the evidence. Only a few tantalising traces remain, including charred plant remains, slender digging sticks from Star Carr and abandoned ground ovens found on the Isle of Colonsay and at other Mesolithic dwelling places.

What is obvious, though, is that while our ancestors were incredibly skilled at obtaining food, it would have been a constant demand requiring the closest attention to seasonal nuances to reduce their vulnerability to shortages resulting from extremes of weather. Today, the alert forager can predict a glut of seeds after drought or a shortage of honey after a cold spring, and I am certain that our ancestors were considerably more alert than that, perhaps in ways that seem inconceivable to us today.

In my own investigation of foraging, for example, I have learned to look out over the forest canopy in December seeking out pockets of oak trees still in leaf. Experience has taught me that in these places I may find frost-free pockets where edible wood blewit fungi may still be picked long after frost has arrested their fruiting elsewhere. Using this strategy to stay alert to the possibility of finding late-fruiting fungi has several times provided me with a meal, even into mid-February. Perhaps the first rule of foraging is to stay alert and to keep an open mind to possibilities.

There is no doubt that our ancestors valued the role of the hunter. Wild game and fish played an essential part in their diet but, unless there was a spiritual taboo, I doubt our ancestors would have passed by any meal, including birds of all sizes, amphibians, snails and other invertebrates, insect life and wild honey. I dare say many today would be shocked if they were to witness the full range of their meat harvest. But what of the vegetable sources of food the forest offered?

My greatest joy in searching any woodland for its wild food potential is not in the food I discover but the way it makes me perceive the forest. Along the way, I always notice and encounter far more than I anticipated. To me, a forest seems like one great organism. Everything living within is integrated in the web of the forest, subtly influenced and manipulated by the forest and subject to its will. I wonder about my forebears: did a Mesolithic hunter look for places where the bilberry is most heavily grazed by deer in November to find a place to lie in ambush? Did that same hunter return to his wife with a still-warm carcass across his shoulders and tell her about the heavy browsing, conscious that she would consider the browsing may be sufficient to stress the plants into fruiting more heavily in the coming year? We shall never know, but these are the sorts of nuances I am always searching for, little seemingly irrelevant clues and hints that can greatly enhance my foraging success.

Now, when we start out in search of wild plant foods, we must accept that not all things are edible and indeed some are poisonous. Fortunately, plant poisoning is relatively uncommon, but never underestimate the danger of eating or even handling some poisonous plants. There are things we need to know to forage safely. Firstly, as we shall see, plant edibility refers only to specific parts of any one plant. If, for example, the root is edible, it does not follow that the leaves are also edible. Indeed, quite the reverse is often the case. It is also important to understand that any plant food should only be eaten when it is in season, that is to say, when they are ripe or in prime condition.

I think of plant foods as sitting on a scale of edibility. At one extreme are those foods that are totally edible while at the other end are the poisonous plants, with the most deadly at the far extremity. Surprisingly, there are relatively few plants at either extreme end of the scale; the majority are to be found scattered somewhere in between.

The most favoured foods are those grouped towards the edible extreme of the scale, although we may sometimes eat plants more towards the middle of the scale for their culinary value or simply because they are more abundant. Usually this requires some specific processing to make them palatable. In times of food shortage, however, human society widens its tolerance of plant foods and may consider consuming plants situated more towards the toxic end of the scale. Processing methods become essential under these circumstances, partly to improve flavour but more importantly to make the food safe to eat. During times of extreme famine, it is not uncommon to find people consuming toxic plants which respond to processing poorly, or sometimes not all. For this reason, we look for precedents to corroborate the use of the most unlikely wild foods. The fact that a plant food has been used in times of famine does not, though, neccessarily recommend it for more frequent consumption.

Most importantly never eat any plant unless you are 100 per cent certain of what you are eating and that it is edible. I would advise learning slowly and thoroughly, one food at a time. Paying great attention to the key diagnostic features, ideally keeping a notebook in which you can sketch each food type, annotating its key features. You do not have to be Picasso, you just need to sketch for your own viewing. While your sketches will prove a useful reminder later, their most important function is to encourage you to focus intently on the minutiae of the details at that moment. Before embarking on the foraging path, remember to go steady at first. It is possible that you may have an allergy to some of the plants. Also, as always with wild foods, avoid excessive consumption of any one food. Moderation is the watchword.

So let's walk through the forest in our ancestors' footsteps and ask a question they would have asked every day. What can I eat here?

EDIBLE LEAVES

At first glance, the forest in late winter can seem bleak and barren, but take a closer look and early stirrings can be found: the race for the sunlight of the coming spring has already begun. For the forager, though, the pickings are slim and, with the trees just waking from their winter sleep, we must look to the plants on the forest floor for food. Many of the early plants above ground are toxic, like dog's mercury, winter heliotrope and snowdrop. Interestingly, the lovely snowdrop is specially adapted to pierce a covering of snow and is equipped with anti-freeze sap. Even the early growing alexanders that sprouts on the edges of coastal woodlands, while edible, is strongly camphorous. At this time in the year it is best to search for moisture, investigating the sides of small streams, damp pockets between root buttresses and boggy areas. Here, some delicious bitter greens can usually be found: young nettles, whose tops can be used as spinach or to make soup and tea. The tastiest grow in shade and are all-green. When they grow in the open, they are more purple in colour and overpoweringly bitter. As the year matures, we shall continue to rely on nettles, even without a cooking pot. The topmost leaves of a nettle can be eaten after a brief wilting in the flame of a campfire to destroy their stings. They are tasty, nourishing and filling.

Next on my list are members of the cabbage family, specifically from the genus *Cardamine*. Hairy bittercress and wood bittercress are found growing in damp, shady places, on the steep sides of narrow streamlets or in humus-rich pockets between the buttress roots of large trees. Delicious, they provide a peppery, slightly bitter, vitamin-rich salad green. These plants can be found year-round but, once they flower, they become more bitter. If you wish, they can easily be grown at home in a pot. If you trim away the flowers before they open, they will grow bushy and less bitter.

As the spring begins in earnest, their larger cousin, lady's smock, will also be found in damp places, though it favours more sunlit locations. This is also a peppery salad plant but is only found through the late winter and early spring. Golden saxifrage is another salad plant which also favours damp ground, particularly spring seepages where it can form a golden yellow-green carpet over very damp ground; it takes advantage of the light in the forest before the forest canopy fills out, blocking out the sun. It is also quite bitter, but not unpleasantly so.

Compared to our modern diet, hunter-gatherers ate many small plants, which, compared to the wonders of modern farming, seem insignificant. Yet the incredibly diverse range of these small wild foods provided a diet rich in vitamins and nutrients. While we have our eyes down, searching the forest floor for edible plants, we will certainly encounter the delicate, emerald-green, clover-like leaves of wood sorrel. These leaves contain oxalic acid, which gives them a tangy apple-peel taste. Oxalic acid is a toxin, so only small quantities of these leaves should be eaten. A few scattered into a salad give it a real zing. Along with these, our ancestors may also have utilised the edible leaves and flowers of wood and marsh violets.

All small, these plants would have been a tonic to our ancestors as they emerged from the nutritionally impoverished winter season. As the year matures, many other edible plants and shoots will exploit the varied light availability under the canopy. We should imagine the delicate nimble fingers of our forebears picking a multitude of nutritious nibbles throughout the year. But it is as the trees set leaves casting shade onto the forest floor that a more substantial supply of wild greens becomes available.

Topping my list of edible leaves are the succulent, refreshing leaves of our lime trees. These are the nearest wild food we have to lettuce and can be used in its place. They are best in the spring when newly emerged but can still be eaten late into the summer when, though more chewy and bitter, they are still remarkably good. Although only available for a very short while in the year, the newly emerged light-green beech leaves are also edible, but they do have tiny hairs on the leaf that can catch slightly in the throat. They are refreshing in a salad but, as soon as the leaves start to darken, they should be promptly taken off the menu.

Hawthorn leaves are also edible when they first emerge and have a slightly nutty flavour. Adding a few to a Waldorf salad provides an extra layer of flavour subtlety.

LEAF TEAS

While the majority of tree leaves are too tough or bitter to eat, many can also be used to make tea. Pine needles, spruce needles and Douglas fir needles have all been used to make teas. To use them outdoors, I simply bring a small billycan of water to the boil then, removing it from the heat, I plunge a branch of fresh needles into the billy and let it steep for a few minutes until a good flavour develops. The needles can then be easily removed. These teas are a real tonic on a cold day and respond well to sweetening with a small amount of honey.

Yew needles, though, must never be used for tea. Indeed doing so may cause fatal poisoning. In one noteworthy case, a 39-year-old man made a wager with a friend on which tea was more toxic, juniper needle or yew needle. Incredibly, the method he chose to determine the answer was to consume yew needle tea. Fortunately for him, he survived, thanks to rapid and effective medical support and prolonged electrostimulation of his heart. Every potential forager should understand that plants and trees are chemical factories capable of producing a range of toxic defences. The toxins they contain can, like yew's, be swift and deadly, while others have more subtle long-term consequences.

If you are wondering, juniper can be used to make a tea, but it is the cones or berries that are usually used, rather than the needles; add one teaspoon of dried, lightly crushed berries per cup. All these delicious teas contain vitamins C and A and are rich in antioxidants. It would be wise, however, to limit their use to occasional due to some of the terpenoid chemicals they contain.

Nettle, raspberry, dewberry, blackberry, bilberry and birch leaves can all be used for tea, too. The leaves are best dried and crumbled and used like conventional tea leaves. Birch twigs are also an option, gathered fresh from trees growing in the open; they tend to be covered in bitter lichen in woodland. Use a handful of fresh twigs per cup. With wild teas, experiment to establish the strength of flavour you prefer and, as always – in moderation.

In nature, life is very competitive. To survive, organisms must have defence strategies. Some evade danger by moving swiftly, but slow-moving organisms must have other defences. Some, like the hedgehog, may be armoured; others, like the toad, may have toxic skin. Plants and trees are anchored in one place and must withstand constant assault by animals, insects, fungi and bacteria. For this reason, they have evolved an ability to generate physical and chemical defences, including a rather sinister range of toxins. The yew is a slow-growing tree that can survive to great age. Unsurprisingly, it is highly toxic.

YEW POISON

The seeds and needles – all parts of the yew tree, in fact, except for the red, fleshy fruit tissue – contain toxic alkaloids, which makes the yew tree one of Britain's most toxic plant species. The flesh of the red fruits, the aril, is sticky and bland but attracts birds who consume the fruit, including the seed. The seed passes through the bird's gut without being broken down, and thus the bird survives and the tree seeds are dispersed. If, however, we humans consume the seed, our stomach acids break down the seed and we are rapidly poisoned. With little food value in the aril and a potentially deadly seed, I consider the yew berry right at the top of the avoid list.

The green parts of the yew tree are equally, dangerously toxic to livestock when browsed. Intriguingly, wild deer have been found browsing on yew seemingly without ill effects, yet it has been shown that deer that browse on yew branches when they have never before eaten them also become poisoned.

So far, there is no known antidote for yew poisoning, which targets the cardiovascular system. It happens very rapidly: symptoms may present within thirty minutes then death occurs within two hours. Should you suspect that yew has been consumed, it is essential to evacuate the patient's stomach as rapidly as possible by inducing vomiting and to immediately seek emergency medical support. Anticipate a rapid pulse rate and ventricular fibrillation, which can result in terminal slowing of the heart and cardiac arrest in diastole. High-voltage discharge and temporary cardiac stimulation might, with luck, bridge the critical phase of poisoning. Symptoms of yew poisoning include nausea, vertigo, abdominal pains, salivation, vomiting, diarrhoea and extremely dilated pupils that do not respond to bright light. The onset of drowsiness, followed by unconsciousness, is frequent. Loss of arterial blood pressure commences as soon as the intoxication begins.

Yew is a stark reminder of the strength of nature's toxins and that they should never be underestimated.

EDIBLE FLOWERS

In my flower calendar, the flowering of the blackthorn defines the arrival of spring. Like tiny white star balls, these brilliant flowers capture the uplifting joy of the sun's returning warmth. If you taste a blackthorn blossom, you will discover a delicate almond flavour, the flavour of prussic acid, a gentle reminder of the chemical potential of plants. A tiny scatter of blackthorn petals in a salad can lend an almond note without causing harm.

Now is a good time to search for ramsons in the moist forest. A wild onion, it is easy to identify and has become very popular with novice foragers for that reason. Very strong-flavoured, it needs careful handling when cooked for risk of overpowering a meal. A pesto from young leaves is an excellent use for ramsons, but my favourite is to scatter a tiny quantity of either unopened flower buds or individual petals from open flowers into a salad. Here, again, it adds intrigue, this time with a delicate garlic pungency.

As the ramsons set seed, pignut stems can be found extending their delicate, wispy leaves. The stem can be followed carefully down into the soil in search of its delicious tuber. These grow eight to ten centimetres underground. As you follow the stem down, it will turn through 90 degrees and then taper down in diameter. Without great care, the root will be broken and the tuber will elude you. If you are careful, you will be rewarded with a knobbly, nutlike tuber about the size of a marble. Squeezed between the fingers and thumb the thin brown skin slips off revealing the cream tuber. Crunchy and peppery like a very mild radish, they are delicious. I am sure our ancestors would have relished them. With experience you can learn to spot them emerging much earlier in the season. I regularly discover them in sunny locations as early as February.

More a food of the past, another tiny root was used in the Mesolithic, the lesser celandine. A member of the buttercup family, its leaves contain a bitter oily toxin that's best avoided. The root is also toxic until cooked. The key to using this plant is to gather it when its leaves turn yellow. Its roots are tiny, similar in size and shape to the tip of a cotton ear bud, but they can be harvested in quantity. Roasted in the embers of a small fire, a process that takes only a few minutes, these taste like potato. Since I first researched the potential use of this plant with my dear departed colleague, Professor Gordon Hillman, charred celandine roots have been found in Mesolithic campfires, testament to their ancient use.

After the celandine harvest, the forest canopy casts the deepest shade on the forest floor. Our ancestors would have been wise to search more open areas, forest clearings where large trees had fallen, alongside streams and rivers where there are more open conditions and sand bars of disturbed soil. For it is in these locations that the best of the edible plants will be found through the summer months.

EDIBLE FRUIT

It is quite probable that large swathes of our broadleaved woodlands used to fall silent through the early summer as foraging parties went elsewhere in search of food. Perhaps only the occasional breaking stick, or glint of sunlight from a flint arrowhead, betrayed the presence of hunting parties stealthily patrolling the forest's dark shadows. As they passed, these parties would have kept a keen eye on the ripening of wild fruits anticipating the most opportune moment to gather these sugary jewels.

Indeed, the first of the fruits to become available in the forest are the most jewel-like of them all, the redcurrants. No other fruit is as beautiful as the redcurrant. When they are ripe, the berries hang like rubies, seemingly glowing when backlit by a beam of summer sunlight. Growing to about 80 centimetres in height, they favour ground where there is moisture. The best place to search for them is in a carr, where alder trees grow in waterlogged soil. The redcurrants will be found beneath the alder but on the better-drained margins. The berries can be eaten raw or cooked.

As the year advances, the forest berries ripen according to local conditions, requiring a forager to periodically monitor their development. Found on acidic soil, bilberries are a flavour powerhouse. However, they will make you tired if consumed raw; they are best cooked with some sugar or dried for use in baking.

Blackcurrants, like their red cousins, also prefer rich, moist soil but are mostly found in rather drier and better-illuminated locations. Overall, they are a more robust plant, growing to 1.5 metres. An easy way to distinguish between the two species when they are not in fruit is to crush and sniff the leaf. Redcurrant leaves have a bland green leaf scent while blackcurrant leaves smell of their fruit.

Wild gooseberries are an introduced species but are forgiven for being such a pure forest treasure. They prefer rich, moist soil and often grow in shade, but they fruit best in dappled shade. Partly because they are uncommon, partly because they grow in locations that seem unlikely to yield fruit, I think of them as the forester's secret wild fruit.

Wild strawberries, by contrast, love a sunny bank and, compared to their cultivated cousins, are a miniature version, smaller than a sugar cube. What they lack in stature they make up for in flavour – a true summer delicacy. I cannot think of another fruit that so raises a smile. I look for strawberries when the horseflies start to bite.

My saddest fruits are our wild plums, damsons and bullace. Sad because I so often find their delicious, abundant fruit rotting on the ground where they have fallen unharvested, reminding me just how disconnected from nature many people are today. Damsons, introduced by the Romans, are divine miniature plums and, rather like the wild strawberry, their smaller size seems in some way to concentrate their flavour. Bullace, our native plum, is similar but rounder in shape; they can vary in colour from yellow through green to black and they ripen later. Both fruits can be used in any way that a plum can: eat them, enjoy them and scatter their stones to encourage more. Picked fully ripe, they are sweet and delicious; picked slightly too early and they have a strong astringence similar to that of the sloe berry, but this can be dealt with, as I will explain.

Of all our wild berries, there are several which fruit with astonishing abundance in most years, but of which little or no use is made today. The most common are the sloe berry and the hawthorn berry. Back when I was researching with Gordon Hillman the wild food that our Mesolithic ancestors might have used, these fruits were of particular interest to us as they are so numerous. Most years, blackthorn berries are to be seen on the branch long after the leaves have fallen, while hawthorn berries paint our hedges and woodland margins scarlet. Here was an obviously abundant supply of food waiting if we could find a way to use them. There were two fundamental issues: how to gather them in quantity from such thorny hosts, and how to make them edible. The harvesting was easily solved: attaching a soft woven basket (a hessian shopping bag could substitute here) to a hooped frame, formed by bending round and binding together two branches at the end of a forked hazel branch. It was now possible to reach up to the berries high in the bushes. The bush could then be struck with a long hazel thumb stick with the free hand.

The results were surprising. First, prodigious amounts of fruit could be gathered in minutes. Second, mostly only the ripe fruit was dislodged by this method. We would go on to experiment with this means, finding that even soft berries such as blackberries could be beaten off the bush; here again the unripe remained firmly attached. Picking became a thing of the past and we could now experiment with truly meaningful quantities of fruit.

Sloe berries are the marble-sized, blue-black fruit of the blackthorn. Of all our wild fruits, they remain one of the most popular to forage, not to eat but to combine with gin to create the ever-popular liqueur, sloe gin. While this liqueur is delicious, I prefer to use them to make a cordial, easily the most flavourful of all our wild cordials. The most frustrating feature of the sloe berry is its tart astringency which renders it inedible when raw. This does disappear after heavy frosting, as the sloe fruit begins to decay, but by that point they are totally unappealing.

The method we came up with to tame the sloe was a purely accidental discovery, and it could not be simpler. Roll each ripe berry firmly between your palms for a few moments and then leave them to rest for a day. I am not certain why this works. Perhaps the rolling liberates enzymes which alter the sloe? Whatever, it transforms the tart sloe into a delicious, plum-like treat. Sometimes it only takes two or three hours to sweeten the sloe by this method, but it will only work once the berry is fully ripe.

The hawthorn berries, called haws, posed a totally different issue. The stone inside the fruit is massive and the fruit flesh beneath the skin only thin. Having tried many methods to remove the stones, bearing in mind that we were using the technology of the pre-ceramic Stone Age, we eventually hit on another absurdly simple method that only required a waterproof container which our ancestors could easily have fashioned from wood or bark. The berries are simply squeezed by hand until the flesh and skin separate from the stones. The stones can then be pulled out from the sticky mass by hand. It is a messy process but very quick. Once the berries are removed, the sticky fruit pulp, rich in pectin, sets hard like a jelly in only a few minutes. This jelly can be eaten fresh or sliced thinly and dried into a fruit leather. This is delicious, with a sweet slight apple flavour, and can be carried as a travelling food. So long as it is kept dry, it will preserve through the winter. The hawthorn berry processed in this way may well have been a seasonal staple food in late Mesolithic Britain.

My favourite berry is the arctic bramble, the most flavour-packed delight I have ever tasted. Sadly, it does not seem to occur in Britain today, although it was once a native plant found on the Isle of Mull and in the Scottish Highlands. Possibly those plants were survivors from more widespread growth during the Late Palaeolithic. Did our Palaeolithic ancestors know this toothsome delight? Very possibly. I remain ever hopeful of finding one when walking in the Highlands.

Until that day, I must content myself with my second-favourite berry, the wild raspberry. Fortunately, this is a widespread British plant, but the casual gaze often misses it, its pink-red berries being mistaken for unripe blackberries. What a wonderful delight it is to find it growing in moist, semi-shaded woodland, particularly on the margins of dewy clearings or streams. It grows on upright stalks, taller than the bramble. Characteristically, the underside of the leaf is a very pale green.

The rambling blackberry or bramble grows horizontally. In sunny locations, it provides copious sweet fruit. The bramble has a habit of infesting disturbed woodland and growing quickly into dense thickets as impenetrable as battlefield barbed wire. In this way, it fulfils a vital role in the forest's development, providing protection for seedlings against the browsing of herbivores. The young shoots and leaf shoots can also be eaten as a salad plant with a delicate coconut flavour.

The dewberry is frequently misidentified as a blackberry. It is less robust than the bramble, and its fruits have fewer drupelets and a cloudier, waxier appearance. They are incredibly delicate and difficult to pick without bursting. Their flavour is more watery and less intense than a blackberry.

Our native apple is the crab apple, small, the size of a ping-pong ball and firm textured. They are too sour to eat raw but can be used for cooking. Rich in pectin, they are a useful ingredient when making jams and jellies from other wild fruits. With an intense apple flavour, they are also good roasted with game.

Apples readily hybridise, so when you encounter any small apple growing in the wild it is worth doing a sourness taste. It is not uncommon to encounter, deep joy, a sweet apple. While the flesh of the apple is delicious, the seeds are poisonous, containing prussic acid. There is one tragic account of a young girl who had developed a taste for the bitter seeds of apples. She collected an egg-cupful to enjoy, sadly with fatal consequence. Apple trees want you to distribute their fruit, so they reward you with the apple, but you must not eat the seed, hence the toxin. Trees are very controlling.

The pear was introduced to Britain at the end of the tenth century. Since then, they have become naturalised and can occasionally be found growing in the wild. They are less common than crab apples, and their small fruits are often ignored, perhaps due to a lack of certainty in identification. Their distinctive fine-toothed leaves turn to gold and then black in the autumn. Wild pears are delicious, a real treasure in the forest, wonderful when cooked in place of cultivated pears. A tarte Tatin of wild pears is a heavenly dessert. Consequently, when I find a wild pear tree, I never forget its location and plan for a future visit.

Although once a familiar fruit, today the medlar passes virtually unrecognised. Related to hawthorn, they have large almost apple-like fruits. However, they are not edible raw until they have been over-ripened, a process called bletting that transforms the toxic parasorbic acid within the fruit to edible sorbic acid. To achieve this, the ripe fruit should be left on the tree until late in October or until after the first hard frost, then picked and stored in a cool, dry, dark place until it begins to decay. Today we can pick the ripe fruits and frost it in the freezer rather than waiting for a natural frost. The frosting bursts cell walls, speeding the chemical changes within the fruit, increasing the sugar content and reducing the acids and tannins.

Bletting can also be used to tame other astringent fruits such as wild serviceberries, sloe berries and rowan berries, although rowan is also best subsequently cooked. Bletted fruits are particularly excellent for making jams.

Rosehips, the fruit of the wild or dog rose, are well known for their high vitamin C content. They last on the bush well into the winter months and remain edible when they are bletted, although I find their flavour reduced by the decay. They are best harvested at the peak of their ripeness. The seeds inside the fruit capsule must be removed as they have hairs that are irritating to the gut lining. This done, they can be eaten raw, dried for future use, cooked into a paste with a little water or dried as a fruit leather. My favourite use is to make a tea by steeping them in boiled water and then to consume the fruit. The hot water bursts the cell walls, liberating their flavour and sweetness.

Whitebeam trees love chalky clay soils, so they are often encountered on chalk downland. They have a mealy berry of no great flavour. The seeds contain hydrogen cyanide so, if the berries are to be eaten raw, they must be picked after frost. Otherwise, having deliberately bletted them, be very careful to spit the seeds out. Gathered earlier, they can be cooked into preserves with sugar. Here again, I would advise deseeding them and producing whitebeam jelly.

One of my favourite sights is a wild service tree in the autumn when its leaves turn red. Each leaf resembles a flame, making the tree look as though it is burning. Its berries are a dull brass colour and slightly smaller than a marble. They are mealy and unappealing until they have been effectively bletted, which transforms them into a delicious tangy fruit. The seeds are toxic and best removed. If you are searching for recipes online, do not confuse this tree for the juneberry (*Amelanchier*), known as serviceberry in North America. They are totally unrelated and have different properties. There are just a few places in Britain where juneberry can be found growing in the wild. Its berries are delicious and can be eaten raw, or dried like raisins.

Sea buckthorn is most often found on the windy coastline of Britain. In our distant Mesolithic past, this dense, thorny tree grew more widely inland and would have been a more familiar sight. The berries are orange, explosively juicy and quite unlike any other of our wild fruits. They grow prolifically around the stems of the tree's branches, often encasing the branch and thus gaining formidable protection from the tree's long thorns. They are one of the richest sources of vitamin C growing wild in the UK and are renowned for their health benefits. While they can be eaten raw, they are best eaten after frost has caused some sweetening by bletting. They must be gingerly scraped from the bush into a vessel with a suitably long tool to avoid the thorns. Thick leather gloves are highly recommended for protection, and be sure to protect your eyes. This is one of the wiriest of our fruit trees, keen to protect its nutrient-packed fruit.

I have always loved the wild cherry. The combination of its beautiful flower and coppery birchlike bark is aesthetically stunning. Though short-lived, they can grow tall in woodland, which puts the fruit totally out of reach. It is only when they are found, shorter, on a woodland margin or hedgerow that there is any hope of beating the birds to the stunning scarlet fruits that are smaller than a cultivated cherry. Be sure to avoid eating the stones, which contain prussic acid.

The bird cherry is most commonly found in Scotland. It is a smaller tree than the wild cherry and produces black, pea-sized berries that are easily picked, although the bird cherry is resistant to being beaten from the bush. The stone is toxic and must not be eaten, and the berry is very bitter, but the flesh of the fruit can be used to make delicious preserves when cooked with sugar. I suspect that this is a berry that would have been used in the production of pemmican in prehistoric times.

Rowan berries are a puzzle. They ripen early in the year, when they are incredibly visible, occurring in large bunches, and yet they are to be found persisting on the tree into winter. One look at them and you feel that you have hit a culinary jackpot. However, while they have been eaten raw, it is probably unwise to do so. They contain cyanogenic glycosides, just as elderberry does. Besides, they are incredibly bitter, so much so that many First Nations in the northern hemisphere considered them inedible. They have been cooked with game, their juice has been used to marinade tough wild meats and they can be used dried to make a tea that, if they are picked after a severe frosting, is quite pleasant. Generally, though, I consider them inedible unless they are combined with other fruits and flavours to make preserves.

The moment I step into the primeval atmosphere of Scotland's Caledonian Forest, I am transported to the north, to the larger boreal forest that it is a part of. This is a stunning landscape, subtly different to the great oak forests to the south. There is a different light through the canopy of needles. This is the land of the capercaillie, pine marten and osprey. The small lochs often hold stands of the true bulrush at their margins, the roots of which, lightly cooked in the embers of a small fire, would have been a delicious, sweet carbohydrate treat for our ancestors.

The forest itself always seems full of promise. The understorey is thin and spaced, comprised of birch, willow and the lovely juniper. Juniper, despite its aroma and famed use to produce gin, has rather fewer food uses. The mature cones which resemble berries can be lightly crushed and used as a seasoning, particularly when cooking grouse or venison. The berries could also be used with fat to make a salve for wounds, or a few could be nibbled to reduce feelings of hunger when travelling.

The ground cover in the Caledonian Forest is a rich and moist pleasant carpet of moss, through which grows a metre-tall mantle of ericaceous plants. This includes Scotland's famous bell heather and cross-leaved heath, whose flowers or leaves can be used to make tea, especially delightful when sweetened with a spoon of heather honey. More significantly, this heathery layer includes three other berry bushes, the blaeberry, the cowberry and the crowberry. Blaeberry is the Scots name for the bilberry, which we have already encountered. Cowberry has a smooth leaf with a dark-green upper surface. The leaf edge curls back towards the underside, which is much lighter in colour and has distinctive pores. Its berries are bright scarlet, they have a pleasant taste but a sour, bitter aftertaste. They can often be found persisting on the bush in winter, though when found under snow they often taste sweeter. Across the boreal forest, they are considered one of the most important wild food harvests. To collect them efficiently, many ingenious devices have been deployed, from salmon backbones, used to delicately comb berries from the bush, to scoops with wire teeth to rake them off. In Siberia, Evenk reindeer herders taught me to make and use an ingenious birch-bark basket. This cunning contraption beats the berries from the bush and then pivots in the hand to catch them as they fall.

Because of their tartness, cowberries are not easy to eat raw, and First Nations in Canada traditionally ate them after prolonged cooking. One of the special qualities of the berries is their ease of preservation: while they can be frozen dry or in water, they can be stored over winter at room temperature by storing the whole berries in containers of water. If sugar is available, they can be preserved by crushing them with sugar or making a jam preserve.

In use, they are best sweetened with sugar, used as a cordial to make a refreshing cold or hot drink or as a preserving juice. A delicious Swedish Christmas dish is peeled pears, first boiled and then preserved in cowberry juice. Across Scandinavia, cowberries are cooked with sugar to make a jam that can be used for dessert or, more often, to accompany game, reindeer, or elk dishes. In Finland, a classic dish is sautéed elk meat and mashed potatoes with a condiment of cowberries, whole or as a jam that's similar to the use of redcurrant jelly or cranberry preserve. Cowberries are only moderately rich in vitamin C, but what they lack in concentration they make up for in quantity.

By contrast, the crowberry is an insipid black fruit with an acrid taste of little promise. As such, few today bother gathering them, and even fewer realise that they are edible. They were widely used across the arctic by First Nations, from the Inuit of the High Arctic to the Sami of Scandinavia. Available in large quantity, they are high in vitamin C, about twice that of the bilberry. Recognising their high water content, they are still used by First Nation hunters as a thirst-quenching trail nibble.

As raw food, crowberries are sweeter once they have been frosted or covered with snow. But they are absolutely at their best when cooked with a little sugar, particularly when used in baking bannocks or pancakes, typical expedition fare, or at home in muffins and cakes. For First Nations, the traditional means of preserving them was as cakes or a fruit leather. They were made by mashing and cooking in bark troughs with heated stones. The fruit mash was spread to dry on bark trays. These fruit preserves were crushed into water for consumption.

One of the mysteries is whether our post-glacial Palaeolithic ancestors also used these berries. Perhaps they added them to fat and dried meat to make a vitamin- and mineral-rich pemmican – a perfect trail food for traversing the vast wilds of Doggerland.

ELDER

Elder is one of those small trees that grows widely yet is hardly given a second thought. It thrives in areas of disturbance, often deep within the shade of the forest. A favourite food of badgers, the trees often grow close to badger setts, where the badgers have inadvertently planted the seeds in their latrine pits. Whenever I find a stand of elder trees close to a stream, I expect to find a badger sett.

A tree with many uses, elder produces copious quantities of fruit, but considerable confusion surrounds the viability of these berries as food. I often come across advice stating that, eaten raw, the berries are only likely to cause a 'minor tummy upset', but I know from the experience of a friend that they can be much more serious. My friend introduced her New Zealand boyfriend to elderberries and, perhaps trying to impress him with her country lore, she mistakenly told him they were sloe berries and totally edible. He, perhaps trying to impress her, munched down heavily on them throughout their long romantic walk. The net result was a burly kiwi rushed to casualty by ambulance and confined for four days' observation. By his own understated description he was very unwell. It must have been a trial of true love, for they are now married.

Why is there confusion regarding the edibility of elder? Well, it seems that there is considerable variability in the toxic potency of the berries on different bushes and between the various elder species. In Britain, we have three elder species growing wild. Elder, our largest tree, grows to ten metres in height and is commonly found across the country. The very uncommon dwarf elder, as its name suggests, is small, growing to just 150 centimetres, and was introduced by early farming communities. The red-berried elder, which is a more recent introduction that has naturalised, grows to four metres and is found mostly in Scotland. Worldwide, red-berried species of elder are the most toxic, but the dwarf elder is also known to produce a high concentration, manifest in its particularly strong unpleasant smell.

Make no mistake: *all* elder species are toxic. They contain toxins in their leaves, bark, seeds and unripe fruit. It is these chemicals that give elder trees their rank, unpleasant odour.

I consider dwarf elder (*Sambucus ebulus*) too toxic and infrequently encountered to bother with. Red-berried elder, (*Sambucus racemosa*) we shall consider in due course. For the moment, let's explore our most common and truly native elder tree (*Sambucus nigra*).

Elderberries are toxic. Raw, it seems that a few can be eaten without noticeable ill effect, but when eaten in quantity they are a problem. The toxins in elder species are cyanogenic glycosides. These are not poisonous in themselves but, when hydrolysed, toxic cyanide and aldehyde are liberated. Hydrolysis can occur when the fruit is crushed or when it is digested in the gut. Just 0.5–3.5mg/kg of body-weight is enough to result in acute cyanide poisoning, resulting in general gastrointestinal upset, nausea, vomiting, diarrhoea (possibly bleeding), general weakness and dizziness.

So why eat elderberries? Despite the seeming risk, they have many nutritional benefits. They contain carbohydrates, amino acids, vita-mins, minerals and beneficial phenolic compounds. Also, given their ubiquity, sheer abundance recommends them as a food resource.

The good news is that it is possible to make them safe to eat. To reduce the risk of poisoning, the fruit should only be gathered when fully ripe, and it must be processed before being eaten, by being dried, heated or fermented. Today, we can dry them and cook them in breads and cakes, or we can boil them with sugar to make cordials, syrups, jams, chutneys and pies; both of these processes greatly enhance their otherwise insipid flavour.

But what might our Mesolithic ancestors have done without a cooking pot? A clue here may come from the careful way in which the more toxic red-berried elder was utilised by First Nations in Western Canada. Historically, the Okanagan people, who live in British Columbia and Washington State, preserved the berries of the red-berried elder through the winter. Their method began with harvesting the berries late in the year, in November, just prior to the first snowfall. They would cut the berry cluster stems at their base so that the berries remained attached in bunches. They would then select a ponderosa pine as a larder tree. The ground at the base of the tree was carefully cleared, and a bed of pine needles was laid down. On this the berry clusters were carefully placed, berries down, stems up. Over them, a thick layer of needles was also placed as a covering. This winter cache would soon be sealed by falling snow. Throughout, the winter people would periodically dig up some berry clusters but they were careful to eat only a small quantity. This process would have possibly prolonged the natural bletting of the fruit, a fermentation that would have helped to reduce the toxins. But given that only a small quantity was eaten, some caution was still needed.

A different method used by the Kwakwaka'wakw people living on the coast of British Columbia involved steaming the berries. Here, again, the berries were harvested in clusters, but the stems were removed prior to the steaming. A ground oven was prepared and, when hot, was lined with the large leaves of skunk cabbage so that any juice would not be lost. The oven was then filled with berries and sealed. After steaming overnight, the hot berries were carefully dried over a slow fire in specially made frames. The resulting cakes enabled long-term preservation of the fruit. A popular food for feasts, the preserved cakes were broken into a dish, softened with water and then worked until they fell apart. These were eaten with eulachon grease, sometimes with salmon berries, which are rather like raspberries. The juice was consumed, but they were careful to spit out both the fruit skins and the seeds, and then drank water to ensure all the seeds were washed out. In this way, we can see heat being used to reduce the toxin but also the avoidance of the seeds where the toxin is concentrated. Personally, I avoid red-berried elder and use our far more common and easily processed native elder.

• • •

These then, with a couple of minor exceptions are the edible fruits to be found in our forests. They are a rich resource indeed, to be enjoyed by the alert and dedicated forager. Many of these fruits also have significant medicinal and health value, rich in vitamins, minerals and antioxidants. But it must be reiterated that there are many poisonous berries that must be avoided, some only mildly so, others potentially lethal.

Learning to differentiate between edible and toxic wild foods would have been an essential lesson in the curriculum of our ancient ancestors' wild schooling. Then, as now, there was only one way to handle this: get to know the plants one by one. At all costs do not follow rhymes and sayings to replace proper identification. A few years ago, I read an article in a respectable British national newspaper that described a journalist's experience attending a survival course. The journalist recounted that he had been taught that 'berries black and blue are good for you, berries red and yellow kill a fellow'. A pity the article had not been fact-checked. Such advice, if it was correctly remembered by the journalist, is potentially lethal. Below is a list of the poisonous wild berries that grow in the UK. It does not include cultivars and garden escapes. Eleven of these berries are black or blue, including the berry of deadly nightshade.

As its name suggests, deadly nightshade (*Atropa belladonna*) is a lethally toxic plant. All parts are poisonous, the most noxious part being the root where, in the case of this plant, the toxins are produced. The least toxic part is the black berry which resembles a very ripe cherry and has a slightly sweet taste. Just two to four berries are considered a potentially lethal dose for a child and a mere handful for an adult, but this of course depends on many factors. The principal toxins are the tropane alkaloids: atropine, scopolamine and hyoscyamine. These inhibit the functioning of both the central and peripheral nervous systems, causing a range of symptoms – face reddening, a severely dry mouth and throat, dilated pupils, blurred vision, staggering and loss of balance, headache, rash, slurred speech, urinary retention, constipation, confusion, a quickening pulse, heart arrhythmia, hallucinations, convulsions, paranoia and cramping. When death occurs, it is usually due to a resulting failure of respiration.

While not all plants are as toxic as deadly nightshade, even mildly toxic plants should not be taken lightly. Coupled with a possibly undiagnosed illness, the result could be disastrous. Sadly, children often make mistakes with berries.

These are the toxic wild berries of Britain, and it would be wise to become familiar with them. Some of the most poisonous are rare and infrequently encountered which increases their danger to the curious.

UK Toxic Berries

Baneberry	Iris
Black bryony	Ivy
Buckthorn	Laurel
Butchers broom	Lily of the valley
Deadly nightshade	Lords and ladies
Dogwood	Mistletoe
Elder	Privet
Elder, Dwarf	Purging buckthorn
Elder, Red	Snowberry
Herb Paris	Spindle
Holly	Tutsan
Honeysuckle	Woody nightshade

EDIBLE NUTS

When it comes to edible nuts, we have fewer choices and less risk associated with toxins. The principal toxic nut to avoid is the introduced species the horse chestnut or conker. Horse chestnut contains a saponin called aesculin, which is neurotoxic at low doses and hemolytic at high doses. It can cause a range of symptoms: spasms and loss of coordination, hypersensitivity, muscle pains, disorientation, diarrhoea, shortness of breath, convulsions and death. Fortunately, it has little flavour to encourage its consumption so deaths from horse chestnut intoxication are extremely rare. The oft-stated advice that it is not dangerous and will only cause an upset tummy, however, is clearly questionable.

Although they are the smallest of our nuts, beechnuts – or beechmast as they are traditionally known – are both delicious and nutritious. Pliny the Elder attests to this in his *Natural History*, when he quotes Cornelius Alexander's account of the use of beechmast by the citizens of Chios to sustain themselves when besieged.

The seed husks are tough with small spines. The easiest way to open them is by placing them in close proximity to a campfire, which causes the tough outer case to open. The nuts are triangular in cross-section. They need to be divested of their shell, which can be peeled off. They can then be eaten raw, toasted or added when baking. They are a wonderful addition to a wild-fruit bannock cooked beside the campfire.

As we have already seen, hazelnuts were a seasonal staple food for our Mesolithic ancestors. Today, unless you live in a part of the country where our native red squirrel still occurs, you are unlikely to enjoy many wild-growing hazelnuts. The non-native grey squirrel nearly always beats humans to the crop, largely because they begin to dine on them prior to their full ripening. This not only interferes with our enjoyment of these delights but also reduces the availability for another species that depends upon them, the dormouse.

Our native hazelnut was smaller than the larger variety that we mostly encounter today. They can be eaten raw, toasted, added to salads, incorporated into sweet baking or, if you are feeling adventurous, cooked Mesolithic-style, in a shallow-scrape ground oven beneath a quick, small stick fire. This is a revelation: an otherwise mouth-drying astringent nut is transformed into a potato-like food.

As I write these words, it is mid-October and I am in the middle of one of those micro-seasons that ruled the social calendar of our Mesolithic ancestors. It is raining acorns in the oak forest. This year's heavy nut yield was totally predictable, the result of drought that stressed the trees through the summer months. Stressed in this way, trees and fungi respond by producing more fruit to increase their genetic survival.

Acorns, however, cannot be eaten raw because they contain too much tannin. To be made edible, the tannin must be removed. The harvesting of acorns would have been rewarding but significantly labour-intensive work for our ancestors. Having observed the strength of the coming nut fall, they would have had to prepare their harvesting equipment, storage and processing baskets. My best guess at their process is as follows.

The first step would have been to identify the trees bearing the best yield, even tasting the acorns to establish which were the least bitter. After this, fallen leaves and branches would have been cleared away from beneath these trees to make the nut-gathering easier. Once the nuts started to fall, more were knocked down with the use of long sticks. The fallen nuts are then easy to gather but are a heavy load of food that would have to be moved to a processing site.

The shelling of acorns is easy if you use a fire. Spreading out a thin layer of hot embers, the acorns can, in batches appropriate to the coal bed, be mixed into the embers and stirred. This heating causes the husk to rupture, enabling the easy removal of the kernel. The kernels would then be lightly broken, not pulped. The nuts would then need to be subject to a washing process to leach out the tannin.

Having seen similar processes employed by Indigenous Australian communities with other toxic plants, I imagine the employment of specialised open-weave baskets designed to be suspended in a slow-flowing stream. The baskets would have to be supervised, lifted if rains caused a flood endangering the precious leaching baskets, and the contents would most likely have been agitated periodically, perhaps daily, for as long as a month. The washed nuts could then be either thoroughly dried for storage as they were, dried and pounded in a wooden mortar to a flour for storage or pulped and cooked in a basket using hot rocks added to the porridge-like mass.

Cooked in this way, acorns taste and smell of rye bread. Really, they are very tasty. Today you can effect a similar process on a smaller scale with an onion bag and immersion in a deep vat of water that is changed regularly. But do not do as my dear friend Gordon Hillman did when trying to speed up the process to meet a filming schedule: he tried to wash them of their tannin by stirring them all night in a hotel bath. The result was the dark-brown staining of a very nice Victorian iron bath. I miss you, Gordon!

The very best first nut for the novice forager is without a doubt the sweet chestnut. An introduced species, this is the chestnut that we love roasted, the one that's sold at outdoor markets at Christmas. They are not related to the toxic horse chestnut, but they bear some similarities, most notably a spiky seedcase and the name chestnut.

And that is where the similarity ends. Sweet chestnuts have the most deeply serrated leaf of any of our trees, and the seed husks have a multitude of tightly packed needle-sharp spines. When they fall to the forest floor, the husk is easily opened by using the inner edges of your shoes to apply downward pressure, causing the husk to rupture. The nuts can then be extracted. They can be eaten raw when ripe, but do this carefully. Using the very point of a knife, pierce the brown outer case of the seed and peel it away, then scrape away the thin bitter coating on the nut and enjoy the fruit of your labour.

They are crunchy and slightly moist but retain that nut astringency that dries the palate. To roast them, cook them in the edge of a campfire in their brown shells and, when they're lightly blackened all over and taking care not to burn yourself, deshell them. This is as above, but it is much easier once they have been roasted. They taste divine with their own unique flavour, and they are no longer astringent. To understand the full potential of the sweet chestnut, you have to visit the chestnut groves in the Cevennes mountains in southwest France. Here, chestnuts are roasted, converted to flour to bake wonderful chestnut loaves, made into liqueurs and soup, cooked in terrines, added to wild game particularly boar dishes, candied as marron glacé and even made into chestnut ice cream.

I also love the versatility of the walnut. It can be cooked with savoury dishes, lending an almost smoky depth to game recipes, and is equally at home in dessert menus, added to fruitcakes or the ever-popular coffee and walnut cake.

The favourite food of the god Jupiter, the walnut was introduced into Britain by the Romans, who grew it for its delicious nuts. It is a beautiful tree with magnificent pinnate leaves, and it is a joy to see it hiding in a hedge or woodland edge. There are several ways to use the nuts. The most obvious is to wait until they are fully ripe before removing the hard-shelled nuts from their green seedcases then cracking open the shell and eating the nut. Freshly harvested, they are moist and wonderful, and, as they dry, they become the nut we are most familiar with.

But there is yet another way to use them that is as surprising as it is delicious, and that is to pickle them. The walnuts must be harvested when they are approaching full size but just before the hard shell begins to form. This will vary from tree to tree but is usually towards the end of June. The whole fruit, including the outer case, is used. To check, slice through a walnut near its base; if you are too late and the shell has started to form, it will show as a distinct light brown layer inside the case. If you judge your walnuts good for pickling, pierce each of them all over with a sharp, pointed tool. I use an awl, but others I know use a knitting needle. It is wise to wear gloves for this job as the juice of the walnut will stain your hands a lasting dark brown; walnut was once used as a hair dye.

The pierced nuts can now be added to a brining solution in a stainless-steel or glass vessel. Leave them to rest in this solution for ten days to two weeks. Although it is not strictly necessary, I prefer to refresh the brine solution after a week. After the brining, remove the walnuts and lay them on a tray to rest. They will gradually turn black, which can take a week or more but can be speeded up by placing them out in the sun on a very sunny day.

Once they are black, prepare a pickling solution and bring it to the boil. At the same time, steam the nuts for five to ten minutes. Using tongs, put the nuts into pickling jars (with vinegar-proof lids) and pour over the pickling solution. Ensure that no pockets of air are trapped, then seal the jars. These should be stored in a cool, dark place for no less than a month before use. They are a delight, a traditional accompaniment to a ploughman's lunch with tasty, mature English cheddar. Truly, the walnut is the food of the gods.

Brining solution
80g of salt per litre of water

Pickling solution
1 litre red wine vinegar
½ tbsp brown sugar or muscovado
3cm of fresh ginger, sliced
15g allspice berries
10g cloves

EDIBLE BARK

Famine is not an experience modern Britons have any experience of. Whether it stalked the land in our distant past, we can only guess. When we consider that we have recorded only 2,000 years of history, it is highly likely that at multiple times in the many preceding millennia our ancestors faced periods of food crisis. Famine mostly results from war and social disorder but can also be caused by extreme weather reducing the normal supply of food. When these factors coincide, a disaster results. The last time any Europeans faced famine was during the Bosnian war of 1992–95. At this time, stinging nettles were sold in the markets and there were reports of people eating fried oak-bark lichen. One does what one must to survive.

It is at such times that the most unusual forest food has been utilised: the bark of trees. My research has revealed that, out of desperation, the bark of many different species of trees have been used: Scots pine, elm, aspen, ash and birch. The most suited are pine, aspen and elm and, of these, the best is pine. It is only the inner bark that can be used for food: in a crisis, thin strips can be cut and boiled and eaten. More usually, the bark is processed into a flour that is added to wheat or rye flour to extend the supply.

In Finland, bark flour called *pettujauho*, known as *pettu*, was utilised very effectively to save lives through famines in the 1690s and 1860s, and during the 1918 civil war. In northern Lapland, I have found many large pine trees still scarred where bark was harvested by Sami reindeer herders during the famine of 1866–68.

In an emergency, bark can be harvested at any time of the year, but it is best harvested in May and June, ideally around midsummer. The bark can be harvested in three ways: by removing a section from a large standing pine; by felling a tree 30–50 years old, chosen for its smooth, branch-free trunk; or, lastly and on a smaller scale, by removing bark from smaller branches or saplings. The process is to carve away the woody outer bark and then to scrape away any of the green inner bark, which is bitter. This leaves the cream-coloured cambium layer. To harvest this efficiently, special sharp spatula-like tools were carved from a branch. With these, the inner bark was peeled in long sheets. These were then processed to remove terpenes, by boiling, by drying or, the most popular method, by carefully toasting the bark.

Once dry, it is dark pink in colour and very brittle. It can now be milled into flour, or pounded in a mortar. In an emergency, it can be put into a clean cloth bag and pounded with a wooden baton against a log. The resulting bark flour is added to porridge or rye flour at a ratio of 30 per cent bark flour to 70 per cent existing grain. It resists the action of yeast so is most effectively used with unleavened breads or crisp bread. While it is a nutritious additive, rich in fibre, iron, manganese, zinc and flavonoids, it should not be used for long periods of time and ideally combined at a lower ratio not exceeding 25 per cent by weight.

* * *

And so our journey through the vegetable foods of the forest ends. It always astonishes me how much there is to be eaten. I have been deliberately conservative in my listing: there are more green foods than I have space for, and I have not even touched on types of edible fungus which, as an organism, falls somewhere between plant and animal.

But even with such a rich diversity of foods, I do not think life was easy for our hunting-gathering ancestors. Several things are apparent to me. To feed themselves, they would have needed to be constantly attentive to the forest. To them, it would have seemed a complex mosaic of microhabitats, in which the availability or other-wise of foods was directly influenced by the availability of sunlight and moisture, by the aspects of slopes and by prevailing winds. They would have needed to monitor the progress of developing fruits, anticipating where best to position their community to take advantage of seasonal gluts. While the berry and nut season seems rich, the concurrence of different fruits ripening simultaneously would have placed enormous strain on a community's ability to make the containers and apparatus necessary for processing and storage of the bounty. Basket-making must have been a highly valued skill in those societies, yet we know virtually nothing about the basketry designs and techniques of the time. Neither do we find any evidence of the ways in which wild foods were stored over winter, when clearly they must have been.

Despite the seeming abundance, there would have been periods in the year when foods that grow outside of the forest or in forest clearings – such as burdock, with its large edible root and green parts, to say nothing of its large and incredibly useful leaves – would have been vital. In the Mesolithic, plant foods were a large part of our ancestor's diet, but shellfish, fish and meat were still, clearly, essential components of our diet that required a wholly different perception of forest resources.

Ivy

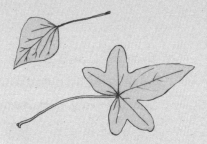

Ancient frescos, mosaics and statues portraying the Roman god Bacchus usually show him wearing an ivy-and-grapevine wreath, and he's generally inebriated. It obviously didn't work for Bacchus, but wearing a wreath of ivy leaves around the head was once thought to prevent drunkenness, a piece of advice known as a prank. What is interesting, though, is the way that the ivy is consistently presented as exerting a stranglehold on the grapevine, twisting tightly to it and choking it off. Could this be the source of the notion that ivy might fight off the effects of the grape?

As a rapacious woody climber that can grow to 30 metres, ivy (*Hedera helix*) has always had a very low reputation but it is not the parasite everyone thinks it is. It has its own independent root system that extracts water and nutrients from the soil, taking nothing from the plants it climbs. Ivy selects older, healthier trees to climb, which naturally already have well-established root systems. Ivy's newer roots are therefore at a different depth from those of the tree it climbs, so the two plants are not in competition for the soil they share. Unable to penetrate bark, it does not damage trees, unless they are already bare or weak or afflicted by pests or disease. The severest threat from ivy to its host is that too much of it will amass on the treetop and block the sunlight; this is solved by pruning the ivy every spring.

Ivy stems have little hairs, trichomes, whose secretions help the plant cling to whatever it's climbing. The maturity of ivy can be assessed by the shapes of its pale veined, glossy leaves:

younger plants' leaves have three to five lobes; those of older plants are heart-shaped and lobeless. For several butterfly species, ivy leaves are a significant food source, as they are for larvae from moths like angle shades, holly blue, small dusty wave and swallow-tailed.

Mature ivy plants produce yellow-green flowers, which blossom in September–November in small, rounded clusters called umbels. Ivy's nectar-rich flowers are ideal for bees, and common wasps and hoverflies join them in collecting ivy nectar and pollen before entering hibernation. Rarer insects, too, are attracted to the plant, such as the golden hoverfly, aka the ivy hoverfly.

In November–January, ivy's black fruits ripen in clusters of round berries. They're poisonous to humans, but invaluable to birds. These berries have a high fat content: according to the RSPB, ivy shares a calorie count with Mars bars, gram for gram. Blackbirds, blackcaps, thrushes and woodpigeons are among those relying on ivy come autumn when other nutrition is becoming scarce.

There is a stubborn belief that rodents are attracted to the ivy itself, but the truth is they're far more interested in your house and kitchen waste. It is true that autumn and winter bring out a multitude of wildlife seeking shelter. As well as birds and insects, ivy offers refuge to bats and small mammals, among them rats and mice.

Another, similar caveat applies to the common worry that ivy will damage buildings. If an old building has structural damage, ivy will probably make it worse. But research conducted by Oxford University for English Heritage determined that ivy provides a 'thermal shield' that reduces 'extremes of temperature and relative humidity', absorbs some air pollutants and protects against frost and salt damage. Ivy can help to preserve old stonework.

In mature oak woodland, dead ivy stems make the perfect wood for friction fire-starting.

Juniper

A ten-metre tree or low-growing, spreading shrub, common juniper (*Juniperus communis*) is an evergreen conifer that flourishes on Britain's chalk lowlands. Its grey-brown bark peels as the tree ages up to 200 years. Its spiky leaves are juniper's year-round principal distinguishing feature: trios of green needles gathered around reddish-brown, ridged twigs. Grey-green on the underside, each needle has a single white strip on its upper surface.

Near the tip of every twig are buds, male and female on separate trees. The small, spherical male ones are yellow by the time their pollen is ready for release. It is carried by the wind to pollinate other trees' green female cones, which develop over the next 18 months into fragrant, fleshy seeds. Small and purple-black when ripe, they resemble blueberries.

These seeds will be devoured and dispersed by birds. Fieldfares, mistle thrushes, ring ouzels and song thrushes all eat the juniper fruit, while firecrests and goldcrests nest amidst the tree's crowded foliage. Where juniper grows in the Northern Uplands, they might even be joined by black grouse. Caterpillars whose foodplant is juniper include the chestnut-coloured carpet moth, the juniper carpet moth and the juniper pug moth.

Exploitation of the juniper revolves around the seeds. Distillers have been using juniper berries in gin since the mid-seventeenth century. By law, a gin's primary flavour must be juniper – if you can't taste the juniper then, legally, you're not drinking gin. Juniper is used as a spice in Scandinavia, where it adds sharpness to meat dishes, notably venison and wild boar, and seasons sauerkraut and cabbage. More recently, it has become a popular element in sauces and liqueurs.

For our distant ancestors, the naturally shedding bark of the juniper's stem was a superior natural tinder as it ignites when slightly damp and can be extinguished for repeated use when fire-lighting. The deadwood also makes for an excellent friction fire hearth. The green wood of branches is hard and springy and in the past has been used to fashion several different styles of fishhooks.

Larch

Introduced to Britain around 1620, the European larch (*Larix decidua*) is native to the Central European Highlands, the only deciduous conifer in Europe. Maturing to 30 metres, from April each year it produces clusters of light-green needles that turn golden-yellow before their autumn fall. Soft, three or four centimetres long, these leaves grow in clumps from rough shoots on the pinkish-amber twigs.

From the tips of these shoots grow larch roses: they look like green, pink or white flowers but are actually scales. In March and April, when pollen is released by the male cones on the shoots' undersides, the wind carries it to the larch roses, which start to ripen. Growing to three or four centimetres, they turn brown before their scales slowly start to open. Winged seeds emerge to be scattered on the wind. Freshly emptied, the cones might remain on the tree stems for years.

Black grouse eat the buds and unripe cones, while birds such as fieldfare, redwing, lesser redpoll and siskin feed on the seeds, as do red squirrels. The case-bearer and larch pug are among the moth species whose caterpillars feast on larch needles, while the larch tortrix eat the cone scales.

When young, a larch is slightly cone-shaped but as it ages, for anything up to 250 years, it broadens and wide vertical crevices appear in its pink-brown bark. Beneath the bark are pale-brown sapwood and reddish-brown heartwood. Larch wood is hard and hardy, largely unaffected by rot and verging on fire-resistant. This robustness explains why it was among the earliest trees to be introduced to Britain specifically for its timber.

Lime

Whether you prefer slices of it in your drink, chunks of it in your chutney or a whiff of it in your detergent, the small, green citrus fruit you call lime has been nowhere near the tree we call lime. That is picked from *Citrus aurantifolia*, a small, Southeast Asian tree. *Tilia*, the lime tree, is a European native, and there are ten or more species of it in the UK today. Most you'll rarely come across without frequenting botanical gardens, but three grow naturally in Britain: broad-leaved, small-leaved and common.

The common lime is a natural hybrid of the small-leaved and broad-leaved. The last of these, *Tilia platyphyllos*, is also known as the large-leaved lime, although the common lime, *Tilia europaea*, has larger leaves; the leaves of *Tilia cordata*, the small-leaved lime, are actually smaller than those of the other two. The common lime, as a hybrid, grows quickly to as much as 40 metres, taller than either of its progenitors. Its fragile leaves are dark-green and heart-shaped and are six to ten centimetres long.

The broad-leaved is the earliest to flower, in June, followed by the small-leaved and, finally, the common in July. The five-petalled yellow-white flowers appear in two- to five-strong clusters, each attached to a pale-green bract. The flowers of the small-leaved are the oddity here because they jut out at random angles.

Soft and a whitish-yellow, lime wood is easily worked and does not warp. Its many applications have included: beanpoles, bowls, carving, cups, fuel, furniture, ladles, piano keys, sounding boards, sticks, including Morris-dancing sticks, and wood-turning. Today, though, they are mainly just ornamental trees.

For our ancestors lime trees may have supplied friction fire-starting sticks, lime being one of the very best choices. But most importantly the bark provided cordage for string, rope and clothing. It could also be folded into baskets and used to thatch shelters, and the wood was easily carved with stone tools to make dug-out canoes. Lime trees were a tree of life wherever they grew.

Common lime

Large-leaved lime

Small-leaved lime

Maple

FIELD MAPLE

The field maple, (*Acer campestre*) is our only native maple and is found growing mostly in southern Britain on chalky soils or with hornbeam and oak. Mostly, it is encountered as a tree of modest size in hedgerows. However, it can grow to 20 metres, and ancient field maples can sometimes be found hiding in the shadows of old established woodland with a girth of over two metres, by which age they are often mistaken for old apple trees. The branches of field maples have a distinctive ridged corky bark. The leaves are modest in size with the distinctive lobed maple shape. The delicacy of the foliage is rather misleading to the quality of the wood. The field maple is the densest, heaviest and hardest maple species to grow in Europe. It is a honey-coloured wood that resists splitting and carves beautifully, holding fine detail. It was a popular wood for turnery and could be turned into boxes with tight-fitting lids. Seasoned, the wood is very hard and polishes well. It was favoured in Tudor times to fashion knife handles, spoons and other hard-working everyday items.

The strength and delicacy of the wood was also harnessed to produce musical instruments, including violins and harps. The Sutton Hoo excavations revealed the Anglo-Saxons' appreciation for field maple. A lyre, an ancient precursor of the harp, was found made from field maple, as were six turned flasks that were elaborately decorated with silver-gilt panels beautifully adorned with animal imagery. Clearly, the Saxons valued field maple and, elevated by their extraordinary craftmanship, we can imagine the intoxicating atmosphere when music spilled from the field maple lyre, accompanied by alcohol poured from the honey-coloured flasks into imported walnut-burr cups, all glinting in the firelight of the feasting hall.

In European folklore, branches of field maple placed around a window would keep bats out of a house. I cannot attest to the field maple's repellent properties, but for insects and birds the field maple is attractive, providing an important piece in the jigsaw of our native biodiversity. It is favoured by ladybirds and hover flies to name but two species.

NORWAY MAPLE

The Norway maple, (*Acer platanoides*) is a native of Eastern and Central Europe, growing to 25 metres. It is believed to have been introduced to the UK during the seventeenth century. It is a tree that is often mistaken for sycamore. It grows in similar locations, suffers damage from bark-stripping by grey squirrels in a similar way and shares the sycamore's leaf infections and infestations. Its leaves, at a casual glance, are almost identical to sycamore, although on closer inspection they are more defined in outline with distinctly spikey lobe ends. They also have a winged seed, like sycamore, but the wings form a wider angle. In the autumn, the

Norway maple's leaves show clearly in the hedgerow or woodland margin as a vivid chrome yellow.

I regard Norway maple highly. It is a lovely wood to carve, it immediately demonstrates a far tighter grain than sycamore. This is a much tougher, springier wood that has found favour for turning and furniture-making. Strong and fine-grained, it can be worked delicately, favouring its use in the manufacture of musical instruments; Stradivarius employed Norway maple for the backs, ribs and necks of his incredible instruments. One of its classic uses was to fashion tool handles that could be beautifully smoothed, and thus would cause no blisters. I have a crooked knife I made with a Norway maple handle, and I have never yet developed a blister using it. The wood can also be used to make bows.

While friction fire can be achieved with Norway maple, there are better alternatives. The leaves, however, can be used to wrap a glowing ember for transportation, as was demonstrated in Ötzi's outfit. The leaves can also be used to wrap food for cooking or as a lining for a ground oven.

Today the Norway maple is not considered a commercial timber due to the damage caused by grey squirrels. It is a great pity, for it is a marvellous wood to work. Personally, this is a feel-good tree for me, with a wonderfully uplifting atmosphere beneath its canopy.

6

SHELTER

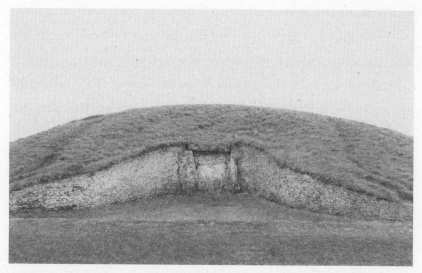

Belas Knap, in Gloucestershire, is a fine example of a Neolithic long barrow.
Thirty-one people were found buried in this ancient house for the dead.

love to walk Britain's coastline. The brisk sea air tightens my exposed skin and provides a rejuvenating charge of natural energy. Watching others walking their dogs, lovers arm-in-arm, or children searching for natural treasures in pools, all seem moved by a mysterious, deep-seated urge to be near to the coastline. The feeling is uplifting.

What many beach visitors do not realise is that they may be walking past remnants of Britain's ancient forests. On many stretches of our coastline at low tide, the sea exposes dark ridges of peat. On careful examination, it is often possible to recognise trees within these peaty masses, even by their species: alder, birch, willow, ash, oak, hazel or yew. Although the scraps of wood look quite fresh, these forest fragments are very ancient, in some places dating to the Bronze Age, in others to the Neolithic, while the oldest connect us directly to the Mesolithic.

Picking out a piece of a soggy waterlogged branch from these forest remains, we connect directly to a landscape where our hunter-gatherers once walked. I cannot help but look at those submerged forests and wonder what life was like when those trees were growing. I am comforted by the thought that some things are timeless, like the experience of walking in woodland in the rain. Now, just as then, a forest deluge never seems to end; long after the clouds have passed the rain continues falling as large, consolidated raindrops from the tree's leaf tips, and with true sylvan egality any creature walking beneath the canopy at such times is equally soaked by the damp understorey vegetation. And this is as it should be, for trees and water are linked by natural association. Forests are meant to be damp places; trees

Edible plants

Some of the wild foods used by our ancestors. In a world without potato, rice and wheat, hazelnuts and acorns were staple foods. Even the tiny roots of lesser celandine could not be overlooked.

Hazelnut

Acorn

Lesser celandine

Sloe berries

Cow berry

Bilberries

Wood sorrel

Rowan

Golden saxifrage

Violet

Wild strawberry

Wild raspberry

Wild cherry

Wild damson

Guelder rose

Medlar

Pignut

Wild apple

Hawthorn berries

Redcurrant

Toxic plants

Many plants protect themselves with toxins. Some of these berries pose a real threat to the unwary or curious, particularly children.

Iris fruit

Holly berries

Butcher's broom berry

Yew berries

Ivy fruit

Herb Paris

Tutsan berry

Spindle fruit

Edible fungi

Fungi play a vital role in the health of the forest, accessing and distributing nutrients between trees. Many form a lifelong association with a host tree. The edible varieties are beneficial to people too; they are rich in B vitamins and contribute significantly as a source of flavouring. These are just some of the edible varieties to be found.

Penny bun

Summer bolete

The bay bolete

Brown birch bolete

Orange birch bolete

Shaggy parasol

Jelly ear

Chicken of the woods

Toxic fungi

Fungus toxins are truly terrifying. These are some of the toxic members of the Amanita/death cap family. All fungus foragers should start by learning to recognise the characteristic features of this family: bulbous base to the stem, sometimes with a sack; a skirt of delicate tissue around the stem just beneath the cap [this can be absent]; white gills that produce white spores; and caps of varying colours with wart-like spots, which can be brushed off or wash off in the rain.

Death cap

Panther cap

Fly agaric

Destroying angel

need water. Their convoluted roots, fallen branches and spongy leaf humus serve to slow the runoff of water and nutrients to the sea.

Staring hard at the peat banks, I can picture a Mesolithic hunter in a damp forest, every leaf sparkling with sun-illuminated rain drops. He sits still and totally silent waiting for a deer to step out from cover in search of grazing now that the rain has passed. Suddenly, a sparrowhawk darts past, just an arm span above his head. The hunter doesn't move; he is invisible in his stillness. But then the picture is lost as the incoming tide drowns out the inspiration of my imagination.

Those forests that our last hunter-gatherers knew were pristine, soft underfoot, with waterlogged mosses and damp humus. They birthed rivulets and streams of sweet water flowing down into rivers that meandered wildly to broad saltmarshes and the sea. It was a time before pollution, effluent and fertiliser; the ecosystem was healthy and in balance; the fluxes in wildlife populations were the heartbeat of the land itself. Was this the highpoint of Britain's biodiversity? Very probably. We can only imagine the abundance of wildlife in such an unaltered natural landscape. Those distant ancestors lived within this natural system and they had no choice but to accept the moisture. Their life was bound by the need to adapt themselves to the environment, learning to read nature's signs well enough that they could foretell seasonal food gluts, and thereby place themselves seasonally in the places where food and other resources were most abundant.

This does not necessarily imply that they were constantly on the move, as is often thought. Nomadic hunting and gathering was perhaps more prevalent during the Late Palaeolithic, when Britain was an arctic wilderness, where the interception of sparse and highly mobile food resources, mostly wild game, demanded a society that was also mobile. The carrying capacity of the land would likely have limited the size of communities, possibly only permitting large gatherings of people in places or seasons when there was a predictable abundance of food.

In the more sylvan world of Mesolithic Britain, life would have been very different. We know that the people made full use of the shoreline with its year-round abundance of shellfish and fish. Is this why we still enjoy the coastline today? Along the coastline of Britain, midden mounds can still be found, the waste heaps of our Mesolithic hunter-gatherers, comprised mostly of the discarded shellfish shells from countless ancient meals. But they also contain the bones from game hunted in the coastal forest.

As the climate warmed, the increasing availability of woodland foods and resources piqued human interest, increasing the diversity of plant foods we ate. It is highly likely that our ancestors found places sufficiently rich in wild foods to require an annual seasonal visit or perhaps even to make them part of their permanent territory. My own experiences working with many of our remaining hunter-gatherer communities suggests that, compared to farming communities, hunter-gatherers are certainly more mobile, but in a localised way within a larger landscape. They change camp for a wide variety of reasons: from beliefs and taboos; after deaths; for reasons of hygiene; to access fresh material resources; or even to resolve a social disagreement. Despite this, they largely remain within the bounds of an established territory, the size of which is closely linked to its capacity for food productivity and where they feel a spiritual connection and cultural identity with the land. As an Evenk reindeer herder once told me, 'The best thing in life is moving on.'

As we have already seen, the first post-glacial explorers of Britain utilised caves for shelter but surely also tents when travelling between cave sites. They must have been capable and skilled travellers to intercept or follow migrating herds of large game. What shape their tents took is difficult to say; certainly some late Palaeolithic sites suggest circular tents enclosing a fire. This is interesting as such a footprint is typically left by the conical tent designs still in use throughout the boreal forest. Called *lavvu* in Lapland, *chums* in Russia, *wigiwama* in Canada, they are conical with a round or slightly oval base and can easily be made smaller or larger as need dictates, providing both living space and working space. In use, the upper two-thirds of these tents hold smoke that slowly escapes from the tent apex. More than just a dwelling, these tents are an important tool. By enclosing the fire, its warmth is more efficiently harnessed and the consumption of firewood is dramatically reduced. The smoky upper space is highly functional. Here, fish and meat and fungi can be strung to dry and preserved by smoking. Skins being prepared for clothing can also be dried here or wedged between the poles and the tent covering.

If we picture such tents being used in Palaeolithic Britain, there are several problems. In Lapland, Siberia and Canada, the long tent poles needed can easily be cut when setting up camp, but it is believed that the forest of Britain at that time did not have much more than dwarf willow and dwarf birch trees growing. So where did the tent poles come from? Were there perhaps sheltered places within the landscape where birch saplings could be found sufficiently sized to construct such a tent and, if so, how were the long, heavy poles transported a long distance by the people as they travelled? Then there is the tent covering. Given the sparsity of the vegetation, it is unlikely that any suitable material other than animal skins or peat could have been employed at that time. Of the tent sites discovered in Europe, a tent three metres in diameter is considered small. So how were people moving the heavy skin tent cover and tent fixtures? In Siberia and Lapland, domesticated reindeer served as pack animals, which they still do in northern Russia, while in Canada dog teams were used to pull sledges.

I believe that part of the answer is to be found in an alternative tent design. Dome-shaped structures were once widely used in the arctic and subarctic, and can be made from flexible willow shrubs, which perhaps were more likely to have been available in sheltered valleys. Just such homes were constructed by the Nunataaġmiut, or Nunamiut, who lived as semi-nomadic caribou hunters in the Brooks Range of Alaska. Their traditional tent was the *itchalik*. This was made from willow saplings, sharpened at their butt ends, pushed into the ground then bent over and lashed together. Where there were willow stands of suitable size, they could be cut as needed. The average itchalik required around twenty poles.

However, in the Brooks Range, as in Palaeolithic Britain, such poles were considered valuable and willow thickets with suitable saplings were at a premium, so it was more usual to carry saplings for the purpose. Saplings three and a half to four metres in length were usually cut in the summer months from good stands of flexible willow. They were cleaned of their bark, then bent into the required shape and dried to retain their curvature.

Once they are dry, willow poles have the advantage of being incredibly lightweight, yet retaining some flexibility and strength, making them a good choice for a tent frame that must withstand high winds. These tent frames were covered with a series of caribou skin sheets, each made from several skins sewn together. The hair was left on the skins for its superb insulation and placed with the fur facing outwards. Over these a second covering was secured, this one made of caribou skins with the hair removed and stitched together with a waterproof seam. Refinements could include a stitched-in window made from translucent bear intestine and a large grizzly bear hide with the fur used as the door covering. In summer, just the outer covering would be used.

This design of tent is extremely practical. The lightweight poles can be more easily transported than long poles, and the shape of the tent is particularly resistant to snowfall and gales. I once built just such a shelter at a Scout site in the Lake District, whilst teaching a Scout troop. That night a gale ripped through the valley. I slept soundly in my dome-shaped mountain tent and arose in the morning to find all the tents that had been pitched around me now vanished. They had not survived the storm, being torn by the wind. It was a strangely eerie experience. I eventually found the missing occupants sleeping inside the large dome-shaped shelter I had constructed from bent willow saplings.

Inside an itchalik, the living space was oval with a width of around three and a half metres and a length of four and half. Unlike a conical shelter, the itchalik has more vertical walls which makes it a more comfortable space to live in. With a height of just one and a half metres, the roof of a dome-shaped shelter is considerably lower than in a cone-shaped tent, making the shelter warmer, enabling it to be heated with just a fat lantern. Drafts were excluded by piling moss against the base of the tent walls, and fragrant flooring was made from small willow branches covered with caribou hides. The itchalik served as home for the Nunamiut for longer than human memory can recall. Archaeological evidence for the use of these tents stretches back 5,000 years, and no doubt they could have been in use for much longer. I suspect that the Nunamiut itchalik provides the most likely recent analogue for the type of tents Britain's Upper Palaeolithic explorers may have used, but it does not solve the question of how they transported their tents, for the Nunamiut didn't employ dogs as pack animals or in teams to pull sledges. Maybe, one day, lucky work with a trowel and brush will reveal evidence for the use of pack animals.

Once the trees in our forests were large enough, I have absolutely no doubt that conical tents would have been employed, as they are quick and easy to construct and astonishingly versatile. Perhaps at first these tents would have been made with a low apex, which resists the wind strongly and is warmer, then later, as the forests thickened, breaking the force of prevailing winds, in taller more upright forms.

By the Mesolithic, it is quite possible that such shelters were being thatched with forest materials, employing sheets of birch bark under a layer of turf or moss or perhaps thatched with reed stems. While birch poles must have been used for such shelters, they are only short-lived. Once alder poles could be obtained, they would have served far better. I build a modern form of just such a shelter while teaching each autumn in Scotland. I use alder poles that were cut and stripped of bark for the purpose more than ten years ago. They are not protected from decay, save for leaning them upright in the forest against a mature spruce tree.

The past 20 years has seen archaeologists unearthing numerous Mesolithic sites with evidence of structures that were most probably dwelling places. Some are indicative of lightweight shelters or tents with no obvious postholes, while others reveal the use of incredibly heavy posts or poles suggesting they were designed for permanent or semi-permanent occupation.

In 2009, a Mesolithic house was discovered at Ronaldsway Airport on the Isle of Man. A thousand years older than any other site yet found on the island, it seems to have burned down accidentally around 10,000 years ago. While this would have been a tragedy for the shelter's inhabitants, it has provided a goldmine of information for modern archaeologists. When organic materials are charred by fire but not consumed, there is a good chance that they will survive, and such is the case with this site, which has revealed several interesting finds. The shelter itself was constructed from hazel and a hawthorn-like wood, possibly hawthorn, apple or whitebeam. The postholes hint at a conical structure, possibly with a skin covering, and it has a sunken floor, a not uncommon feature in Mesolithic dwellings.

One of the rare opportunities afforded by such a find is to peer into a frozen moment of Mesolithic domestic life, to discover items not normally found and even to see how the home was organised. From the details of the investigation so far published, it appears that remains of charred hazel within the shelter may indicate that it incorporated a wooden floor. Perhaps the sunken floor provided an insulating airgap to reduce conductive heat loss. Inside the shelter, stone tools were found including a uniquely shaped sandstone. Was this a cherished abrading stone for working hides, a tool for processing plant foods or perhaps for shaping wood? I hope that in time further research will reveal more.

Flint tools were knapped at specific locations within the shelter, with hammer stones possibly contained in a willow or poplar basket. Over 10,000 hazelnut fragments were recovered. From their concentration within the structure, it appears that they were stored at the edge of the shelter either side of the entrance, while they were being processed near the centre of the shelter where there is possible evidence of cooking pits or waste pits. Certainly, cooking hazelnuts in a shallow pit oven in a shelter would be an excellent way to control the process and to avoid burning the nuts. It might also have been the cause of the conflagration.

Despite these incredible insights, perhaps more significant is the testimony the shelter bears to the courage and seafaring capability of our Mesolithic hunter-gatherers. The Isle of Man became separated from the mainland 85,000 years ago and now sits in the middle of the Irish Sea, roughly 50 kilometres from the United Kingdom. Reaching the island by boat today is not a journey to undertake lightly, let alone 10,000 years ago in a prehistoric dugout canoe or skin boat.

Hazelnuts are a regular find in the Mesolithic houses that have been found. They were a staple food of the age, and the discarded shells were quite naturally cast into the fire for disposal. Given the vast quantity being consumed, many were only charred and thus survive for modern analysis. They can be very useful in determining the age of sites by radiocarbon dating.

At Howick in Northumberland, hazelnut dating was invaluable in the investigation of a more permanent Mesolithic shelter found on the coast, eight kilometres north of Alnwick. Initially constructed around 9,800 years ago, the Howick Mesolithic house is Europe's best-understood Mesolithic dwelling. It was also equipped with a sunken floor, on which 50 centimetres of deposits had accumulated. Investigation revealed a succession of hearths and many postholes, stake-hole slots and pits excavated from the floor surface. Amongst a large quantity of flint objects and some ochre, organic remains were also found, including charred hazelnut and acorn shells and burned bone fragments from birds, wild pig, fox (which tastes somewhat like lamb) and a canid, either a domestic dog or a wolf.

Even though at the time of its occupation the shoreline would have been several hundred metres away, there were also marine mollusc remains. Associated with the shelter but outside of its periphery were numerous contemporaneous shallow pits or scoops with evidence of charcoal, charred hazelnuts, marine shells and occasional flints, precisely the remains that would be expected from shallow-pit cooking of hazelnuts, the de-husking of acorns and the quick-fire cooking of shellfish. Most importantly, the shelter shows signs of continuous use for a prolonged period of time, being regularly maintained, repaired and occupied either permanently or semi-permanently for a period of a hundred years or longer. This suggests its locality was once a place where the natural resources aligned to allow for permanent settlement.

While we are fortunate to have these astonishing archaeological sites, they can only be a tiny fragment of what once was. Many smaller shelters made of small poles will be invisible to us today, their traces long since ploughed away. It is worth considering the great southern forests of lime and elm trees that once grew. If ever there was a landscape where our native inhabitants could have been described as forest people, it would surely have been there.

The advantage of the lime and elm forest is that the bark of both species can be stripped off in large sheets to easily make durable weatherproof roofing. This would have enabled our ancestors to make huts from small saplings, either as conical designs or perhaps more efficiently as dome-shaped structures considerably larger than the Nunamiut dome tent. The inner bark of both trees would also have supplied the necessary cordage to tie such shelters together.

If we could step back in time, we would likely locate such an encampment by following the scent of woodsmoke. If we did, we might find a cluster of neat dome-shaped huts providing enviable dryness after an autumn shower, smoke rising peacefully from the roof holes of the huts. Delayed by the oppressive moisture, the smoke might be lingering in the stunning canopy of golden lime and yellow elm leaves. A woman perhaps gazes for a moment at the lingering smoke in the canopy and thinks of the menfolk who have gone to smoke out a bees' nest in a large lime tree; she is hoping to taste honey later. Stooping to enter the doorway of a shelter, she takes firewood inside where her daughter is sitting on a woven bark mat gently bouncing an infant in her arms. Inside, the shelter is snug and tidy. Woven bark bags are suspended from the walls. At the edge of the hearth are flint chippings illuminated by a single beam of light that is piercing the fug of thin smoke from the hearth that fills the roof of the shelter. There, on racks, are strips of dried meat and fresh fish neatly and uniformly butchered and beginning to dry.

In many ways, these people are wealthy. Their trees are incredibly giving, providing long fibres that can be woven into a wide variety of baskets for storage or for load-carrying, conceivably even for clothing. The wood of the lime is light and easily carved with flint tools and, when dry, can be smoothed with rough sandpaper like elm leaves. In spring its leaves are succulent and edible and, later in the year when they are too leathery to eat, they can be used to wrap food or to line ground ovens.

When this extended family decides it is time to move camp, the remains of the camp will be difficult to locate in perhaps as little as two years. The shelters will have collapsed and decayed and returned to the forest, and the enduring debitage from stone-tool production will be lost under a blanket of fresh humus and leaf litter. Silently the flint chippings will wait. Perhaps 10,000 seasons hence, an amateur archaeologist will spot them excavated by a burrowing rabbit. Try as they might, though, no archaeologist will be able to unravel a trace of that ephemeral encampment.

Although Britons had successfully lived by hunting and gathering for just short of a million years, and more recently for 9,000 years immediately following the melting of the glaciers, the tradition was not to last. Everything changed, and rapidly so, with the sudden arrival of a new people on the British coast.

Sometime around 4000 BC, give or take 50 years, a party of people walked over to a large tree, in a corner of England we know today as Kent. They were excited. In their hands they carried a new type of tool. Slotted into carefully shaped ash handles were flint axes, which were valuable, invested with hours of extensive polishing on sandstone. Their hearts were resonating to the buzz of the revolutionary new idea that drove their industry and their purpose. These people would have arrived by boat, probably entering the Thames estuary, and hugging its southern bank before turning south into the meandering River Medway.

It is possible that this forest at the foot of the North Downs, near to present day Ayelsford, between the towns of Chatham and Maidstone, had been found by a small scouting party the previous year, who came searching for a quiet place with little evidence of the native hunter gatherers; a place where they could establish a community without interference or opposition.

The site was close to several sarsens which, as fixed points in the landscape, may have had established trails leading to them. They saw promise in this land: the chalk soil would supply the quantity of high-quality flint they would need for their tools; there was good hunting and fishing; most importantly, the calciferous soil supported a wealth of plant life. Selecting a suitable tree in a sunny location one of them stepped forward and swung the new tool at the perfect bark of a mature tree. As the stone blade bit into the wood the chopping sound rang out resonating through the forest. It heralded the moment when our relationship with the forest was forever changed. These people wanted the tree to build a house but more importantly they wanted to clear the forest.

This was the beginning of the Neolithic. These people were pioneers in the vanguard of a new way of living that had in the past thousand years spread like wildfire across Europe from the Aegean. They were farmers. From this moment forward, people would cease to adapt themselves to the land. Instead the land would be altered to suit humanity. The forest would be felled to create fields to grow crops and to raise livestock. Farms would anchor people within the landscape, and they would set up permanent homes. Families could and indeed would need to be large to supply the labour force necessary for this new experiment in food production. It was in truth a fine balance. Many hands were needed to exploit the farming potential of the land, but should the crops fail then the people would swiftly experience famine. The stage was set for land ownership, land jealousy, conflicts over property, divine explanations for crop failures and, ultimately, for warfare.

The Neolithic farmers were industrious and well organised. They would remodel the British landscape, leaving behind long barrows, houses for the departed that were a visible proclamation of a community's long and continuing connection to their land. Even today we are still awed by many of their megalithic monuments.

There must have come a day when a party of native hunter-gatherers spied smoke rising from the growing clearing and new settlement, close to a sarsen that today we call the White Horse Stone. When they approached it, they would have found a long rectangular house, probably with planked walls and a doorway in the middle of its western side. Seventeen and a half metres long and nearly seven metres in width, it would have seemed huge, dwarfing any shelter they had ever seen before. This design would have been as alien to them as were the aspirations of these strangers.

We can only imagine now the discussions in the hunter-gatherer communities about these new people and the way they lived. Whether there was antagonism or conflict between these opposing interests is not known. But soon the hunter-gatherers would find themselves competing for edible plants with the grazing livestock of the new arrivals, who lived in larger communities, were organised to provide communal support and who were determined to convert the land to their way of living.

The sound of that first axe blow would echo again in later centuries as similar cultural changes were inflicted on the African, American and Australian continents. In the archaeological record, subsistence by hunting and gathering in Britain and in fact the world ends abruptly after the arrival of farmers. We can only guess at the possible cultural loss, but we are more certain from bone analysis that a healthy, diverse, mineral- and fish-rich diet was replaced by a limited, nutrient-depauperate diet, the consequences of which we still struggle with today.

Unpolished flint axes are prone to chipping, which can in turn lead to their snapping in half. The stone axes of those Neolithic farmers were polished to endure the heavy work of felling mature trees. In many ways they symbolised their philosophy, for these farmers were thinking in an enduring way, and their use of materials would reflect that. Their houses were fashioned from heavy ash and oak supports. Woodworking would advance with the new toolkit, which included pottery that facilitated boiling, parching, storage, fermentation and brewing.

Exchanging the sylvan innocence of the hunter-gatherer for the labour of agriculture, they could store food and free themselves of time that would otherwise have to be expended chasing game. In this freed time, ideas were born, experiments conducted, discoveries made, new technologies evolved. It was the beginning of an industrial revolution; it was the time of Stonehenge. People quickly forgot the hunters and gatherers, whose lives now seemed so woefully primitive in comparison to the achievements and new ideas alive in the age of farming. In time they would vanish, genetically subsumed into the larger farming population. But somewhere deep in each of us there is a vestige of our hunter-gatherer forebears.

It is perhaps worth considering that since those first fields were created, Britain has lost most of its forest. Today woodland represents only 13 per cent of British land area, the least of any European country. It seems that in our excitement with our technological advances we lost sight of important things. When we weaken or remove a forest, it allows rainwater to accelerate with gleeful abandon, eroding and washing away the life-supporting soil, carrying it into the sea as a sediment to choke littoral and marine life.

But today it is not just soil that is washed into the oceans: driven by the maxim that 'dilution is the solution to pollution', chemical pollutants, fertilisers, effluent and rubbish are all washed into the ocean on an unimaginable scale. I wonder, if we could gift a family of hunter-gatherers a glimpse of the future that is our present day, would they be alarmed by the lack of trees? Would they think that our cleanest rivers are dead by comparison to the world they knew? The problem is so great that plastic waste can now be tracked by satellite. Recently, the seasonal movement of plastic waste rafts in our oceans were plotted, but more alarmingly plastic waste could also be detected pouring en masse from rivers at a rate estimated at 14 million tons per year.

Medlar

There is a sequence in Romeo and Juliet in which Mercutio pulls off at least five medlar-based smutty puns in six lines. Shakespeare's sense of humour has not travelled well through 400 years, and the medlar fruit itself has entirely lost its popularity, but the sixteenth-century audience would immediately have grasped not just the innuendos but also the routine slur in the implication that Juliet will be rotten before she is 'ripe'.

Medlar (*Mespilus germanica*) fruit looks like a very small (five-centimetre) brown-skinned apple. It has a hollow indent that gave medlar its medieval slang name, 'open'ers' or, as Mercutio recalls, 'open-arse'. The flesh inside tastes very bitter and feels hard until it ripens; it is basically inedible. Unfortunately, its ripening requires exposure to frost or simply being kept in storage for two–three weeks. This managed rotting is called bletting. The fruit loses some of its tartness, becoming acid-sweet, but the insides also lose their resemblance to fresh fruit, becoming a slightly putrid purple pulp. It can then, eyes closed, be eaten raw, or used for curds or jellies. Bletted medlar was reputedly popular with a glass of port after a meal.

Originating in Southwest Asia and probably brought to Britain by the Romans, medlar's popularity lasted from medieval times to the Victorian age, but it is now vanishingly rare. The small tree is sometimes still found in the woodlands of warm areas of the country, and it can be identified by its long, narrow and wrinkly leaves and, once they bloom in May, its five-petalled white flowers with their white stamens, yellow-brown anthers and, behind the flower, conspicuous green sepals.

Oak

Our ancient forests truly must have been a wildlife paradise, for this was the heyday of Britain's biodiversity. Fortunately, we can still find natural mature oak forest to provide us with a glimpse of prehistory. Oak trees (*Quercus*) support more forms of life than any other native trees. They are the living embodiment of biodiversity.

In marked contrast to the managed forests of today, the prehistoric landscape would have been wonderfully chaotic. Deadwood was commonplace, fulfilling a vital stabilising role in the forest ecosystem, generating fresh, nutrient-rich soil that retained moisture, reducing run-off and the associated erosion of soil and nutrient. Hollows in dying and dead trees provided homes for birds and mammals, including bats, while beetle larvae furnished an essential supply of food.

Almost from the moment an oak tree sets leaf, it is attacked by aphids, caterpillars, leaf-mining moths and gall wasps. Invertebrates are often overlooked, but they are one of the most important components of the forest ecosystem, contributing more than just a source of food. Earthworms and woodlice break down forest litter, while dung beetles live up to their title. The tiniest springtails break down dead vegetation and fungi. Many British invertebrates are saproxylic, dependent on dead or decaying wood for all or part of their lifecycle. This includes many species that today are considered rare or endangered, such as the stag beetle. Fungi too abound in deadwood, efficiently recycling the nutrients and redistributing them within the forest biome.

Oak trees have always been a valued resource to human society. For our hunter-gatherers, acorns were a source of sustenance. Despite their high tannin content, which had to be laboriously leached out, the seasonal glut of acorns would have been a vital staple food. Oak wood is hard, strong and durable which suits it for the construction of buildings, machines and ships. In the Bronze Age, dug-out canoes were made from oak. But it also has an even-textured grain, allowing it to be carved and turned. Oak wood has a distinctive scent that results from its high tannin content. Oak bark used to be harvested as a by-product of timber extraction and used in the tanning of leather.

Perhaps it is in barrel-making that all of oak's qualities are best exhibited. The cooper utilises the strength of the oak and its ability to be precisely carved to form staves, which are water-tight yet flexible enough to be bent into the bow of the barrel. Such barrels are still highly valued, not just for their durability but also for the tannin flavour that they lend to the various alco-hols brewed or aged within them.

The Oak Trees

While today we are blessed with numerous species of oak, only two are native trees: the English or pedunculate oak and the sessile oak. Both can be found virtually anywhere in the United Kingdom. They are long-lived trees and can form high forest. The main difference between the two is the habitat they favour, although they can frequently be found in prox-imity and readily hybridise.

The English oak (*Quercus robur*) is often found in coppice woodland and ancient wood pasture. It grows best in heavy, fertile soils and, while it prefers well-drained soil, it can cope with a moderate degree of waterlogging and can often be found beside fens or small forest ponds. It does not do well on acid soils or shallow soils, particularly overlaying limestone. It is most easily identified by its seed; its acorns have a long narrow stalk, while its leaves have a very short stem.

The sessile oak (*Quercus petraea*) grows best in well-drained soil, but it can cope with acidic soils. Being more shade-tolerant than the English oak, it can regenerate within the shade of existing forest. Its acorns do not have a stalk (sessile means 'stalkless fruit'), but its leaves have a longer stalk than the English oak's. The sessile oak also grows well in uplands, even above 300 metres and on difficult rocky terrain (*petraea* means 'rocky places') or where there is heavy rainfall.

The English oak is the national tree of England and, fittingly, the sessile oak is the national tree for Wales and Northern Ireland.

At least one of our old oak trees is named by an event that occurred within its shadow. In the winter of 1810, the shady bole of an old oak tree was momentarily illuminated by sparks from a flint igniting the powder in the pan of a keeper's gun. The resulting report signalled the demise of a female sea eagle that had fastened on the head of a deer, plucking at its eyes. It was the death of one of the few remaining British sea eagles, our largest bird of prey, which was once common but would, just a few years later, be extinct within England and, soon after, Scotland. Ever since that gun fired, that old English oak has been known as the Eagle Oak. It is still growing and can be found in the New Forest's Knightwood Inclosure, where today it is sheltered by tall conifers. Beside it, yew trees grow which seem to crown the ancient oak with an evergreen canopy.

Consider the changes that our oldest trees have lived through and the stories they might tell. Unlike our brief lives, measured in mere decades, the span of ancient oak trees is counted in centuries. Our oldest oaks began their lives ten centuries ago, long before machinery, fertiliser or pesticides. When they first reached up from their acorn in search of life-giving sunlight, the air was pure and clean, and the forest was abuzz with insects and the music of myriad songbirds. The raindrops that nourished their cohort of seedlings was sweet and pure. The night sky was truly dark, filled with the clear twinkle of a billion stars and the bright, mysterious trails of shooting stars. Today their leaves are greasy with pollutants, the rain that falls is acidic, the night sky glows a sickly orange, and there are fewer songbirds singing in their branches.

Pear

The common pear (*Pyrus communis*) is a large, deciduous tree that thrives in light, deep soils. The domed crowns of the tallest-known pear trees have hit heights up to 20 metres, but most will stop nearer 12. Their grey-brown bark sports square tessellations, while their branches boast spiky twigs and white hairs swathe their small, pointed buds. In the autumn, their spring growth of light-green serrated oval leaves turns gold before blackening and dropping to the ground. Before it falls, this foliage offers ideal long-term habitats for caterpillars as they await pupation.

The end of March sees the pear tree come into flower. The three-centimetre blooms comprise five white petals, five yellow stigmas and abundant red anthers. If bees and flies don't report for pollination duties, the flowers are able to self-pollinate. The anthers turn a darker red or purple once their pollen is released.

The fruits grow on long stalks with no hollows at their stems, which means that they start out as thin, cylindrical tubes that then broaden at the base. As they ripen, the pears become sweet-tasting and golden-coloured. Blackbirds and thrushes are at the front of the queue for this fruit, and the seed they subsequently disperse will swiftly find more deep, light soil to germinate in and repeat the whole process; discarding a pear core will have much the same result.

Pearwood tends not to splinter or warp and it doesn't hold on to smells, flavours or colours, and seasoned and sanded pearwood is as smooth as plastic, so it's long been used to make kitchen utensils like wooden spoons. It also has a good pedigree as material for the making of woodwind instruments.

Pine

As the climate warmed around 10,500 BP, Britain's birch forests were gradually being shaded out by a new species. Like birch, the Scots pine is a pioneer species, able to colonise open ground, particularly where the soil is sandy, poor or acidic or a loamy gravel. This enables them to grow on heathland and in mountainous country.

Their winged seeds can disperse 150 metres from the parent tree. Pine seedlings that rooted in the shade of the birch forest had the advantage of being evergreen, which enabled them to photosynthesise throughout the year, storing energy to speed their development. At winter's end, the pine had a head start. Before the birches had set leaf, pine saplings were using the thin shade of the birch canopy as the perfect nursery. They gained height in search of sunlight at the expense of establishing girth, easily outmatching the birches, growing through the canopy to heights up to 35 metres. For several thousand years, as the birch declined, magnificent pine forests grew to cover an estimated 1.5 million hectares of northwest Europe. Today, just 16,000 remain in Britain. Pine woodland is now commercial forestry, where the trees grow like soldiers in dark, straight, narrow-rowed plantations.

The Scots pine is a climax species, a long-lived tree that can create a stable, long-lasting forest ecosystem, supporting a wide diversity of life. This is the country of the Scottish wildcat, pine marten, red squirrel, capercaillie, black grouse, crossbill and parrot crossbill and crested tit, and of many unique plants, like twinflower, wintergreen, cowberry, creeping lady's tresses and bilberry. The fungi are special, too. Try an autumn search for the dusky-toothed *Sarcodon* fungi, the greenfoot tooth fungus, or *Sparassis crispa* (cauliflower fungus). Each is a marvel of nature.

The Scots pine (*Pinus sylvatica*)

The Scots pine is Britain's only native pine tree. It has slightly twisted blue-green needles that occur in pairs.

Male flowers develop at the bases of shoots and are most obvious when heavy in bright-yellow pollen. The female flowers emerge in May at the ends of shoots. Initially, they are red, turning purple once they have been pollinated, after which they develop into cones. Pine cones have scales with a central ridge and do not open to release their seeds until the spring of their second year.

The Scots pine is most readily identified by its bark. Near the crown of a mature tree, it is orange and seems to glow in the warm light of sunrise and sunset. The bark at the base is rust-coloured, tending to dark purple and deeply fissured. In the cracks of the bark, all manner of creatures can be found, including chrysalises.

Pine is one of the most productive trees grown in forestry, growing tall, straight and fast. The timber is strong, light and durable, properties that perfectly suited making ships' masts and spars. It's also tight-grained, responding well to sharp tools, which has long recommended it for furniture- and cabinet-making. Sawn into long planks, it was perfect for flooring. Now, it is frequently pulped and processed into composite construction materials or for paper-making.

A strip of the inner bark can be used as an emergency plaster, bound in place with string or natural fibre. To seal damage to its bark, pine produces resin, which can be collected for a variety of uses. Our ancestors may have used it fresh as chewing gum. Heated and reinforced with charcoal or beeswax, they could have made a thermoplastic glue; such glue has been found adhering to Stone Age arrow shafts. Once heated to drive off the turpentine, the resin becomes rosin, the friction-increasing substance that musicians rub onto the bowstrings of violins. Without it, stringed instruments would be barely functional.

The thin rootlets of pine have long been used for basketry and binding. In northern Finland, where pine forests dominate, the roots are also used to make strong ropes.

Now mostly forgotten, one of pine's most important products is tar. Extracted from short sections of wood from the thicker roots by dry distillation, the black, oil-like tar was invaluable for waterproofing and preserving ships' ropes, hulls and deck caulking. It was also the original ski wax for wooden skis and is still in use today.

Pine is an exceptionally good fire-starter. A split section of wood from an old stump is highly resinous shaved into a feather stick for fire-starting, arguably the best of its kind. However, pine wood burns with a resinous soot that disqualifies it as a wood for the smoking of meat and fish. Pine can also be used for friction fire-lighting.

Poplar

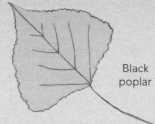

Black poplar   White poplar

Poplar trees are not native to the British Isles, but they have been here a very long time, predominantly in two common species: aspen and white poplar. That list would once have listed a third species, but the black poplar is now rare and declining, with only about 7,000 still growing in the wild, chiefly around Cheshire, East Anglia, Shropshire and Somerset. There are also several hybrid varieties, including natural hybrids like the grey poplar, which is very common here.

Now naturalised in Britain, the white poplar (*Populus alba*) originates in central and southern Asia and central and southern Europe. Most often seen near water, it is a deciduous broadleaf tree which thrives in moist conditions. Able to grow in coastal sand where it can withstand ocean winds, white poplar may be planted as a windbreak.

Its pale-grey bark displays lenticels, lines of black diamond-shaped pores. At first growth, the twigs are densely covered in white hairs but, as they age past a year or two, they lose the hairs and become very rough and knobbly. The shoots and buds are also covered in white hairs, and the undersides of the leaves as well, so the white poplar really earns its name in summer. A greyness to the dark-green upper leaf adds to the overall snowy impression. White poplar has variable leaf shapes, but the most common have five lobes and flattened stems. The leaves are irregularly serrated.

The dioecious white poplar flowers in spring. Male flowers bloom as red catkins, which drop from the tree when they've released their pollen. Once wind-pollinated, the yellow-green female catkins on other trees then turn green before releasing clouds of fluffy white seeds to carpet the woodland floor in late summer.

The tree's caterpillar catalogue includes dingy shear, pink-barred sallow, poplar grey, puss moth, sallow kitten and yellow-line quaker moths, which all feed on its leaves, while birds eat the seeds.

White poplar's timber is soft and very light and therefore quite flimsy, but its bark has anti-inflammatory properties so it has been used in the treatment of gout and back pain. It can also be used as an antiseptic, poplar bark poultices being useful for treating infected wounds.

While white poplar can grow to 20 metres in height, it is dwarfed by the black poplar (*Populus nigra*), where it still exists, which may reach 30 metres. Black poplar's faintly balsam-scented leaves are heart-shaped, and they attract the caterpillars of figure of eight, hornet, poplar hawk and wood leopard moths.

Timber from black poplar is also pale and soft but its natural suppleness and resistance to shock have made it more useful, for example when making cartwheels. It is also naturally fire-resistant and was once a standard material for floorboards. Hybrid black poplar timber is now used in artificial limbs, bowls, pallets, shelving and wine cases.

Black poplar's numbers are now so low in this country that black poplars are no longer able to pollinate each other. Instead, the remaining 600 female trees tend to become pollinated by other species, resulting in further hybrids and the quickening decline in genuine black poplars. The species is now at the head of the list of endangered timber trees in Britain.

7

FIBRE

Common nettle fibres were an important source of fibre for
our ancestors, used to attach Mesolithic arrowheads
and to fashion fabric in the Bronze Age.

S ometimes archaeological discoveries seem like messages sent to us from the past. In 1998, just such a discovery was made. It was John Lorimer, an amateur archaeologist and beachcomber with a keen eye, who made the discovery. At first his discovery was not taken seriously, but eventually his dogmatic persistence paid off, with local archaeologists asking Maisie Taylor, Britain's foremost specialist in ancient wood, to investigate. She immediately recognised the importance of John's discovery and the urgency needed to investigate it before it was lost to the relentless power of the tides. There, on the quiet Holme Beach on the North Norfolk coast, a ring of oak timbers could be seen at low tide protruding from the beach. Today, we know the site by the rather misleading name Seahenge, for the structure was neither a henge nor was it originally sited beside the sea; rising sea levels have brought the coastline inland since its construction.

Prehistoric archaeology usually consists of two-dimensional footprints of long-vanished structures. For a three-dimensional construction to have survived was indeed a gift from the deep past, one that has enabled us to peer more deeply into the lives of our early Bronze Age ancestors. The structure comprised a ring of 55 oak timbers, each split in two and erected so that their sides abutted tightly, with the bark surface facing outwards and the brighter split surface facing inwards. It is known that the trees were all cut at the same period, during the spring and summer of the year 2049 BC. Study of the cut marks left by the builders' axes has revealed that around 50 different axes were used during the fabrication.

Originally erected on the edge of a saltmarsh on the margin of fields and woodland and estimated to have stood three metres in height, it would have been an impressive sight. There was a narrow entrance formed from a forked trunk, which faced towards the direction of the rising sun and was closed with another wooden timber. Contained centrally within the monumental palisade was the immense bulk of a large mature oak tree, inverted so that the root buttresses pointed towards the sky like the fingers of a mighty hand. It was a rarity in archaeology, a mystery, and a unique prehistoric treasure of national importance.

While there can be no certainty over the construction's purpose, it is not impossible that it had a very mundane purpose. Considering its location away from habitation and the considerable effort it took to construct, though, perhaps the most convincing explanation is that it was used for excarnation rites. If so, the upturned roots may once have cradled a body while nature divested the skeleton of its flesh and skin. The tight tall walls were strong enough to protect the departed from unwelcome disturbance by scavenging game. Seen from the outside, it would have resembled the trunk of a mighty tree. Did the upturned tree represent the return of the body to the earth? Is it a reminder of a time when the people of Britain felt a deeper more spiritual connection with the forest and when trees symbolised a direct connection with Mother Earth?

When the massive central trunk was eventually raised from the silt using a nylon sling and the hydraulic power of a mechanical digger, evidence of how it was originally manoeuvred was revealed. The base of the bole had been neatly cut flat. Close to its base edge, deep notches had been cut with the intervening wood perforated to create two shackle-like points of attachment to tow the log into place. Here, still tied securely around the timber and passing through the tow-holes, was the rope that had been used to haul it into place. Rope and string from prehistory is a rare find indeed, but to find a rope still in situ was an incredible revelation. When examined, it proved to have been fashioned from long pliant stems of honeysuckle, hitherto a plant which had never been recorded in such use and which, while known to be useful for basketry, was not considered a suitable material for use as a rope, any such knowledge having long since passed from our cultural memory. Three long honeysuckle stems between 15 and 22 millimetres in diameter had been twisted clockwise to partially separate and loosen the fibres in much the way a withe is made pliant. The three Z twisted stems had then been plied together anti-clockwise, taking advantage of the honeysuckle's natural inclination to twist anti-clockwise, to form a strong S lay rope.

Fibres that can be twisted or plaited together are an essential part of the human toolkit. They are so ubiquitous that we do not even notice them. As you read these words, the chances are high that you are wearing clothing woven from the twisted fibres of cotton. With vines, plant tendrils, animal sinews and even heart strings pointing the way, we can be certain that the discovery of string-making was made in the depths of our deep past and would have been a revolutionary discovery. I have no difficulty picturing an ancestor discovering the method and showing it to a friend who, with a 'let me try that', also smiled in astonishment at the incredible discovery.

The oldest suggestion for the use of string by humans is provided by ancient seashells dated to between 135,000 and 100,000 years old. Perforated seashells found in the Skhul Cave, Mount Carmel, Israel, seem likely to have been strung together to create a necklace, as do similar finds made at Oued Djebbana in Algeria. The most compelling of such finds, however, comes from the Qafzeh Cave in Israel, where ten 90,000-year-old bittersweet shells (*Glycymeris insubrica*) have been found that were deliberately perforated and exhibit wear patterns that affirm that they were suspended next to each other on a string, necklace fashion. Some also show ochre staining. Whether the cord was made from plant or animal fibres is not known.

Fifteen thousand years ago, it is likely that Britain's post-glacial explorers were making their ropes and cords by cutting strips in spirals from animal skins. While these can and have been made from scraped but unprocessed rawhide or from tanned and softened skins, it is the durable and immensely strong rawhide that through history has found the widest use, lashing frames of sledges together, lacing snowshoe frames, plaited into lassoes and stitched into dog sled whips. Despite the easy availability of modern fibres, skin ropes are still commonly employed across the Arctic, particularly sealskin cords that are resistant to freezing. Sinews, the natural threads from animal tendons, are also still widely used. They are the original sewing thread used to sew hide clothing together and would have been critical to the lives of our ancestors, enabling them to make life-preserving clothing for cold climates. Sinews are so enduring that the stitching can outlast the leather it connects.

But I have always thought that plant fibres must also have been used at this time. The classic insulation for reindeer-skin footwear is sedge fibre, classically harvested from the white sedge (*Carex canescens*) and other similar species. The stems of these sedges are dried in the summer and then softened by pounding. The soft fibres are then stored for transport by twisting or plaiting them into a thick rope which is rolled into a flat, round coil and neatly wrapped. A good handful of these fibres in each boot serves as a sock. I have used them and can attest to not only their warmth but also their practicality. Should they become wet, they can be removed, and the loosely separated fibres quickly dried in the warm air beside a campfire. Given the choice between sedge and modern socks I would unhesitatingly choose the sedge, as modern socks cannot be dried anywhere near as quickly.

In 2020, the discovery of the earliest example of human-made cord yet found was announced, dated to between 41,000 and 52,000 years old. At the Abri du Maras rock shelter in a valley that feeds into the River Ardèche, close to its confluence with the River Rhône in southern France, archaeologists had been finding twisted plant fibres in association with Neanderthal tool waste. The breakthrough came when a tiny fibre fragment was found adhering to the underside of a Levallois flake, the Levallois technique of flint-working being, in many ways, the calling card of Neanderthal people. Only 6.2 milli-metres long and 0.5 millimetres in diameter It was not until it was examined with a microscope that it was recognised as a fragment of cord. Made from the inner bark fibres of a conifer, possibly juniper or pine, it was fashioned from three strands of fibres that had each been twisted to an S lay, before being combined by twisting anti-clockwise into a Z lay three-ply cord. It was a delicate masterpiece.

While modern, machine-made ropes are three-ply, most cords made by hand are two-ply as it is considerably easier and less time-consuming to manage two strands rather than three. If greater strength is required, a cord of thicker diameter is made. When cord is made three-ply, it is indicative of a specific need for greater strength and/or better handling qualities, a special cord that will require extra care in its manufacture. That this cord was fashioned three-ply at such a small size is interesting. What might such a twine have been used for? Attaching a spear point? Binding together scarf-jointed spear components? Making a fishing line or snare?

Clearly, it represents high skill in cordage-making and an alert-ness to the fibres available from trees and plants. Cordage allowed our forebears to make ropes, nets, mats, rain capes, bags and baskets. Such technology made possible a host of other technical possibilities: the construction of tents and buildings, rafts, boats and bridges. The storage of plant foods through winter months and the manufacture of rough clothing eventually led to the invention of the loom and the development of the fine-fibre clothing we take for granted today. Obviously, the fibres available for use were dependent on what was available in the nearest forest. As the forests of Britain flourished in the warming climate of the Holocene, so too the range of fibres available became more diverse.

I have had a keen interest in this subject since I was 15. In 1979, I was exploring the types of natural cordage to be found in my local woodland. I had already learned to make strong ropes from the naturally shedding outer bark of the clematis vines, and had experimented with willow bark and bramble bark. But I was particularly interested in the use of stinging nettle fibres. I had read about nettles being made into cloth in Nepal and about their use on the northwest coast of North America, but all the information I could access in that pre-internet age described a mass-production method which involved beating large quantities of nettles into fibres or retting them slowly in troughs of water.

Retting allows natural microbial breakdown of unwanted pectins and lignins, allowing separation of the individual bast fibres from each other and from the outer bark. It is most dramatically effective with lime bark, which undergoes a miraculous transformation. When lifted from its soaking places, the bark smells as bad as fermentation can smell, but now the inner bark can be stripped out in masses of fine strips. It is rather like a magician producing an endless string of silk handkerchiefs. Rinsed of any adhering dirt or gloopy pectin and strung over a branch to dry, these strips lose their unpleasant odour.

If water is not available, a similar process involves laying out the bark in a damp meadow and utilising the effect of dew and sunlight to the same end. Lime bark ropes were still in regular use until the end of the twentieth century.

Back to the nettle. My initial exploration revealed its great strength, but I felt certain there was a more expedient method to access its fibres. I can clearly remember the August day when I sat down with a pile of nettles to solve the riddle. The method I came up with is simple and easy.

First select the correct nettle. Nettles can be used as soon as they approach maturity but are at their very best, being at their tallest, when they have flowered. I prefer nettles grown in shade, in large clumps. September is my favourite month for nettle collection.

I do not use gloves but given the greater prevalence of allergies suggest the use of leather gloves when initially handling the nettles. I cut the nettles just above their very woody stem base. The leaves are stripped with one firm pass through my hand. This also strips away the stings. Done correctly, firmly and with confidence, I am not stung. I will process a batch of nettles in this way.

The stripped stems can now be flattened, either by squeezing or gentle pounding with a light baton or, as I prefer, simply by walking on them. The aim here is to flatten the stems, not to totally soften them. The stem can now easily be split and opened out flat.

I take the flat nettle and at its centre bend it such that the woody inner fibres snap and begin to separate from the outer fibres. The nettle can now be hung over the first finger of my non-dominant hand. Placing the thumb lightly on the fibres, I pull on the end of the nettle with my dominant hand. The result is that the pith lifts off from the fibres as they pass beneath the thumb. By turning the nettle around and repeating the process in the opposite direction, all of the pith is removed, leaving a strap of four fibres still attached at the leaf nodes. These can easily be stripped into four separate ribbons.

These fibres can be used immediately for simple tasks, but if they are to be made into cord will require drying first. It is a general principle when using vegetable fibres for cordage that they need to be dried prior to use. They shrink substantially as they dry. If made into cordage before this drying has occurred, the tight twist of the cord will open and loosen as the fibres dry. Dried nettle fibres shrink considerably in thickness and become wiry.

Once dry, they can be used as they are for simple cord, although to make the very finest cord they will need to have the green outer bark removed. I achieve this by simply pulling the fibres against the top edge of my thumb nail. Straight away the green outer pith will separate, taking the lumpy nodes with it and leaving softer, silvery cream fibres behind. While this final step makes for the best resulting threads and is essential for weaving nettle cloth, it is not always necessary. I only take this last step when it is needed. The prepared fibres can now be twisted into cordage, which I'll come to in a moment.

A similar method can be used to process fibres from goat willow-herb stems. These should be gathered after flowering when they are at their strongest and the outer bark is no longer green, but a golden colour. Strip away the leaves, flatten and split open the stems as for nettle. With this plant, the pith does not separate so readily as the nettle. In this case, break the woody pith every six centimetres down the length of the stem and carefully strip each small section away from the outer pith. These fibres are ready for immediate use for simple cord.

OUTER BARK: CLEMATIS

Clematis vines are commonly found in the woodland of chalk down-land. Through the winter, their downy seedheads look like smoke on the trees. As children, we would swing Tarzan-like on these vines across chasms, until eventually one unlucky Tarzan snapped the vine and discovered gravity. The outer bark of old woody clematis vines sheds a dusty bark in long fibrous strips. These can easily be gathered in quantity and used directly to make very strong ropes.

TREE BAST: WILLOW, LIME, ELM, OAK, SWEET CHESTNUT, PINE, JUNIPER, SPRUCE AND CEDAR

The bast of some trees can be used to make a variety of cord. As we have discovered already, Neanderthal people were making incredibly fine cord from inner bark. To do so they would have had to be very selective of their materials, only using very thin inner bark from branches. The thicker the stem the thicker the inner bark.

Inner bark is best harvested when the tree is growing vigorously, from the late spring and through the summer. Through the winter months it adheres tenaciously to the wood making harvesting impractical.

With the conifer species, the usual method is to cut or scrape away the outer bark and then to strip off the inner bark from the woody stem. This is also the method for harvesting thin willow bark from thin saplings. More generally for willow, lime and elm, the outer and inner bark are stripped off together and then the inner bark is removed.

This can be achieved in either of two ways. The outer bark can be scored through and broken away from the more flexible inner bark.

This is a quick and expedient way to obtain fibres for immediate use. Alternatively, the bark can be left to soak for four to six weeks in a slow-flowing stream or pond. The bark needs periodic monitoring during this period as retting can occur faster if the weather is very warm. Left too long, the bast degrades. The strongest fibres are also said to be obtained by retting in seawater.

More woody parts of trees can also be used to produce cordage. In Finnish Lapland, strong ropes are twisted together from long, spreading roots of a Scots pine. Whether such ropes were made here is not known, but withes made from pliable stems and branches were used extensively. Withes are a virtually forgotten binding material and are, in many ways, the natural wire of the forest. Two hundred years ago, the woodmen supplying the bakeries of London with firewood sent faggots of it into the capital daily, each bound in withes. These withes were in turn returned to be used again until they were finally too dry and brittle. A withe is a thin shoot or branch that has been twisted to loosen its fibres, but without being weakened. They can be made in sizes from small withes of four millimetres in diameter to withes made from whole pine saplings. Mostly, though, they are made from shoots 1–1.5 centimetres in diameter. They are made from a variety of trees, mostly from, hazel, willow, birch, dogwood, pine, spruce, yew and wayfaring tree. Birch is particularly well suited, being usable even in sub-zero winter weather, when other trees become too brittle.

To make a withe, a straight shoot or side branch is selected, ideally of sufficient length and an even diameter through most of its length. Any side shoots are removed with a knife, being careful not to score the central stem. I prefer always to give this shoot a good firm pull before twisting it. Then, starting at the thin end, the shoot should be twisted in one direction until its fibres are heard to crack; at this point, the twist needs to be progressively worked down the stem to its base. The classic way to achieve this is to bend the tip of the shoot so that it forms a Z-shaped handle, somewhat like a bicycle pedal. This provides the leverage necessary to twist the fibres. As the fibres loosen, the twisting is moved down the shoot, always twisting the yet untwisted fibres. Where there are knots or branch joints, the stem will resist twisting, and here particular care must be taken to avoid stalling, overtwisting and breaking the fibres above the resistant wood. Once the shoot is fully twisted, it can be cleanly cut free.

Good withes, particularly those of hazel, can be split in half to make excellent bindings. Withes can answer a thousand binding needs, from making impromptu livestock collars to lashing together a log raft or fashioning an emergency rope. They are also very durable. I have built rustic structures from withes that have survived intact for more than ten years.

LAYING UP TWO PLY CORDAGE

Step One

Take a quantity of fibres half the thickness of the desired cord. Fold them in half and twist both ends in the same direction. The fibres will begin to ply together at the bend; continue until you have five centimetres twisted up.

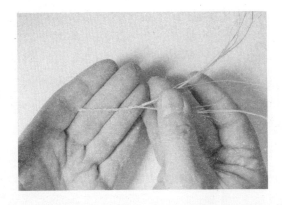

Step Two

With my left hand I pinch the place where the fibres are twisting together. With my right hand I keep the two strands apart and prepare to roll them between my thumb and middle finger.

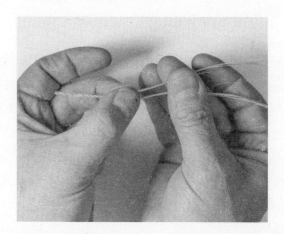

Step Three

Keeping the strands apart I roll them simultaneously between the thumb and middle finger of my right hand. At the end of the roll, I clamp them tight so that they do not untwist.

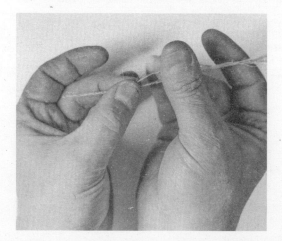

Step Four

Release the fibres clamped in the left hand and the strands will twist together. Encourage the twist so that it is tight. Repeat steps two and three until one strand becomes thinner or shorter than its pair, at which time new fibres will need to be introduced.

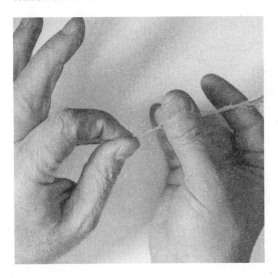

Step Five

Lay new fibres alongside the thinner or shorter strand. Match the number of strands introduced to perfectly maintain equality of thickness in the two strands.

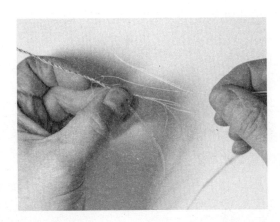

Step Six

Twist the new fibres into the strand that they are being added to. Now continue as before, continuing to add fibres when necessary, until the length of cord desired has been made.

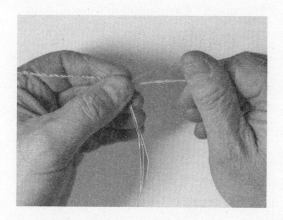

LAYING UP TWINE CORD OR ROPE

Making a cord from fibres requires the fibres to be twisted into strands. These strands are then combined by twisting them together in the opposite direction, counter-twisting. Cord-making does not require very long fibres to make the strands. Of far greater importance is that the twisted strands are of even and equal diameter throughout their length. Should one strand be thinner, it will tend to wrap around the thick strand. At such a point, the even loading of the strands is lost and the rope will only be as strong as the thickest strand. The strands can be twisted up as the cord is being laid or by making long strands prior to laying up.

To make a simple two-ply cord of prepared nettle fibres, start by deciding on the diameter of the cord you intend to produce. Select enough fibres to twist into half of that diameter. Bending the fibres in half at their middle, twist both ends in the same direction;

they will quickly begin to twist together where they are bent over. With care, they can be made to twist together very tightly. With five centimetres twisted in this way, hold the point where the two strands separate from the twisted section, by pinching the junction between the thumb and first finger of your non-dominant hand. Now, between the thumb and first finger of your dominant hand, roll the two separate strands in the direction you began with. They must be rolled simultaneously and to equal amounts but must be kept parallel and separate from each other. At the end of the roll, clamp the twisted fibres between your dominant thumb and first fingers so that they cannot unravel. Now release the cord from beneath your non-dominant hand's pinch grip. The cord should twist together of its own accord. By all means encourage it. When it will no longer twist, pinch it again and repeat the process. Thus, cord is produced.

Naturally, the strands will run short, but usually well before that one or other of the strands will diminish in diameter. It is critical that this be prevented. To achieve this, lay sufficient fresh fibres alongside the strand to maintain the strand diameter and simply twist them in. This constant adding of fresh fibres is the secret to making cordage. Keep going in this way until you have made a metre of cord. It is wise at this point to wrap the growing cord around two short, pencil-thick twigs as a bobbin. This will allow the lengthening cord to rotate as the cord is made. Failing to do this will result in the cord unravelling because it could not rotate as it requires.

Once we start to look closely, the forest is filled with fibres which can be twisted together. This method is the fundamental principle of cord-making, whether producing strong ropes for ship rigging or the finest threads for weaving. But to witness the potential of the method we need to return to Bronze Age East Anglia.

At Must Farm Quarry, near Whittlesey in Cambridgeshire, the remains of a Bronze Age settlement were excavated in 2015–16. Dubbed 'Britain's Pompeii', the site revealed three round wooden houses that had been destroyed by fire around 3,000 years ago. Constructed on stilts, the burning buildings collapsed into water below, which extinguished the flames and provided unique preservation conditions for the organic remains, which were both charred and protected in waterlogged silt. Their state of conservation was exceptional. Cooking pots were recovered still containing the food that was being cooked on the day of the fire, along with the valued tools and utensils of everyday life. The site revealed the community's extensive use of trees chosen for specific purposes, such as a wheel on an oak axle and a palisade made of ash.

Most impressive of all, though, were the textiles recovered. Along with fibres in all stages of preparation were twined fabrics of lime bark, nettle fibres processed into yarns and wound onto stick bobbins, and flax fibres loom-woven into cloth with a thread count of 26 threads per centimetre, which approximates the weave of low-cost modern bed sheets. The finds are black or dark brown in colour today, a result of their charring and having sat in silt for three millennia, but in their day these fabrics would have been attractively light-coloured or maybe even colourfully dyed. The quality of the workmanship is exceptional, demonstrating the expertise and care of the weavers.

The find has once again transformed our understanding of the past and the capabilities of our ancestors, raising the distinct possibility that people at that time, along with growing flax to produce linen, were also clothing themselves in fabrics home-woven from nettle fibre and lime bark, literally the fabric of British woodland.

Rowan

In Old Norse, it was *reynir*, the tree that saved the god Thor from being swept away in a river. Irish myths saw it brought from the Land of Promise by a race of giants. Welsh tradition accused it of being the wood used for Christ's cross. Its Celtic name was *fid na ndruad*, 'wizard's tree' or 'tree of the druid', part of its long association with all things witchy. In medieval England, the 'witch-wiggan' supposedly offered protection against witchcraft: its berries are red, believed to be a good colour for fighting evil, plus each berry has minuscule markings slightly resembling a pentagram. More prosaically, eighteenth-century binomial nomenclature allocated *Sorbus aucuparia*, the Latin root of *sorbus* meaning 'reddish-brown'; more poetically, in 1804, it picked up the name 'rowan'.

Rowan's long, oval leaf is pinnate and comprises serrated leaflets in between five and eight pairs and one terminal leaflet; it grows from hairy, purple leaf buds on hairy twigs. Blackbirds, fieldfare, mistle thrushes, redstarts, redwings, song thrushes and waxwings plus caterpillars of the apple fruit moth all feed on rowan's berries, which achieve that vivid scarlet in July, having developed through the summer from dense clusters of cream-white, five-petalled flowers. Botanists class these berries as pomes. Unlike other pomes such as apples and pears, rowan berries must be cooked before humans can eat them, and they make a rather sour jam.

Rowan prospers in fairly high altitudes, growing to 15 metres and living for 200 years. Its pale-brown wood is not durable, despite being hard and strong, but can be used in craftwork, engraving, furniture and turnery. Often the first tree to flower in mature woodland, rowan is a beautiful wood to carve green. Thicker sections have a deep-brown heartwood that was favoured by rustic carpenters in the production of spoons and utensils.

Sea Buckthorn

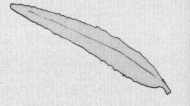

A member of the olive family, this is a deciduous shrub native to Britain's seaside areas, where the soil inevitably has a high salt content. If the level of salts in soil water is too high, water may flow *from* plant roots back into the soil, but sea buckthorn (*Hippophae rhamnoides*) is salt-tolerant, and it flourishes harmlessly near the shore.

However, problems have arisen with it being planted further inland, where it has revealed itself to be highly invasive at the expense of other plants. It grows shoots at its base and spreads itself by using these basal shoots as suckers and quickly creates impassable, thorny thickets that prevent other plants from growing; planted too near a house or driveway, it will do damage. The only answer is to remove it entirely, but volunteers on the Sefton Coast in Merseyside reported that it grew faster than they could dig it up.

In early spring, clusters of very small orange-brown flowers blossom on some trees; these are the males. Even tinier, yellow-green females grow among the thorns on other sea buckthorns. After pollination, long, narrow green leaves with silvery scales emerge alongside the appearance of orange berries on the female plants. These berries contain ten times more vitamin C than oranges, but their juice freezes at such low temperatures that a household freezer won't freeze it. Raw, they are so extremely bitter that even most birds will not eat them, but they can be used in jams or liquors after a few days' bletting. It requires a commercial operation, for both industrial freezing and harvesting (although sea buckthorn can be harvested manually, it's a prickly affair). The juice from the fruit then needs to be filtered to remove sand and sweetened with honey to make a traditional winter tonic.

Spindle

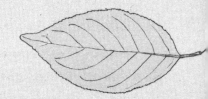

The spindle tree (*Euonymus europaeus*) grows to a height of nine metres. It is a delicate tree found in the hedgerow or woodland margin on chalky or well-drained soil. Although a delicate tree, it can live for one hundred years or more. It is most easily spotted in the autumn when its beautiful, delicate russet leaves offset its stunning fruit that resembles vivid pink popcorn. In the spring it develops flowers which are very similar to dogwood, with four creamy-white petals in a cross arrangement. However, on close inspection the petals are narrower and more delicate than dogwood, and the leaves bear no resemblance at all. The twigs and young branches have a dark-green bark adorned with four corky stripes.

The colourful fruits often attract the attention of children, but both the leaves and the fruit are toxic to humans and should not be consumed.

In woodland, spindle is an indicator of ancient woodland and plays an important role in the forest biodiversity, providing food for many small creatures, including the caterpillars of spindle ermine moths and holly blue butterflies. In folklore, it was widely considered to be a lucky tree.

It derives its name from one of its traditional uses. Spindle wood is very hard and stiff and riven splints can be sharpened to an enduring point. It also polishes well. These properties made it an obvious choice for making spindles for spinning whorls, as well as for knitting needles and sometimes skewers. However, it does not naturally produce many long straight stems, so was not considered to be an arrow wood. The hardness of spindle suited it for turning small items such as lace bobbins. It was also used to produce darning needles and combs. The baked and powdered seeds were employed to kill lice.

Strong and slender and often producing wide Y-shaped branching stems, spindle was a natural choice to fashion crutches. The same properties in more slender stems were used to make handles for anglers' landing nets, the hoop completed with a bent section of another wood. Its strength meant it could be used to make a delicate, lightweight but robust handle to meet the needs of the most ambitious angler.

Today little use is made of spindle, but high-quality artists charcoal can be made from small twigs. To do this, prune fresh twigs, usually of pencil to little finger thickness, although the size can be larger to suit your artistic preference. The sticks can be used fresh or can be dried prior to charring. Fill a tin container that has a secure fitting lid with the sticks. An old coffee bean tin is ideal for this. In the lid, punch a 2-3mm hole. Place the tin into the centre of a small campfire, with the hole uppermost and visible. As the wood inside is baked, fumes will be seen flowing from the hole, which will eventually ignite as a small flame. When the flame goes out, remove the tin from the fire using tongs or gloves to prevent burns and plug the hole with a sharpened stick to exclude air. Leave the tin to totally cool, then open and remove the charcoal sticks. Willow and alder can also be used.

Spruce

It is now thought that the festively ornamental Norway spruce (*Picea abies*) may have been native to the UK between 130,000 and 115,000 BP, the last interglacial period. Perhaps it was, but it was then absent until it was (re)introduced in 1548, so it is classed as non-native. That has not stopped this evergreen conifer from becoming a common sight in Britain's forests, gardens and parks. The other spruce brought to Britain, the Sitka in 1831, has been planted on an industrial scale to meet the needs of the timber production industry. Estimates of the lifespan of both species range from 200 to 1,000 years; perhaps this uncertainty is down to the fact that they are almost exclusively planted expressly to be cut down in their primes.

Fast-growing and very straight, Norway spruces are triangle-shaped with a pointed crown some 40 metres up, if they are fully grown. The bark of the young tree looks smooth and grey-brown, but is actually rough to the touch; by the time the tree reaches maturity at around 80 years old, that bark has become a cracked and dark purple-brown. They are, of course, among the most familiar trees in the country: perhaps not everyone recognises their orange shoots, and maybe some are unacquainted with their profusion of downward-hanging, reddish-brown cones, as much as 20 centimetres long, with overlapping rounded-diamond scales, but everybody knows their leaves, short, dark-green needles, pointed but not sharp, and with fine bands of white on their undersides. They give out a sweet smell, even when stacked up in garden centres late in the year.

The Sitka spruce is even taller than the Norway, able to reach at least 55 metres, and sometimes as much as 80. Its needles are straight and flattened, and these needles are actually sharp. The bands on their undersides are blue-white.

In May, once yellow clusters of male stamens have released their pollen, the Norway's upright, oval female flowers turn from red to green and start to swell as they develop into those red-brown cones. The flowers of the Sitka are similar, but the cones that the pollinated females become are pale-green cylinders. By the time they ripen in autumn, they will be a crinkled creamy brown, still pale, and they will ultimately release seeds that are winged and small. The flowers and cones of both species favour the upper parts of the tree, but will appear further down too, especially on the Norway, whose reddish cones are a seasonal treat for red squirrels.

The Norway likes a bit of personal space, but the Sitka spruces will grow very close together, their thick foliage discouraging other plants from growing around them. This dense canopy provides a natural umbrella for larger mammals to shelter beneath, and offers good nesting for lots of small birds – the coal tit, crossbill, siskin and tree creeper among them – and to larger birds of prey, like the threatened goshawk. The Norway is host to beetles, hoverflies and weevils and the inevitable range of moth caterpillars, barred red, cloaked pug, dwarf pug and spruce carpet among them.

For two hundred years or so in Britain, the strong, pale-cream timber of the Norway spruce has been used to make boxes, flooring, furniture, joists, paper and rafters. The Sitka spruce accounts for 50 per cent of commercial plantations, and its extremely versatile timber is used for making boats and ships, pallets, packing boxes and much more. While not ideal, spruce can be used for friction fire lighting.

Sweet Chestnut

This is another classic British tree that is actually non-native to these shores. The sweet chestnut (*Castanea sativa*) has been around since classical times, and the Romans were great enthusiasts for its fruit, which is both sweet and a nut. It is also vitamin- and mineral-rich with a high starch content, and, famously, it can be roasted on an open fire. It can be used to make biscuits, cakes, nut roasts, soup, stuffing and a variety of confectionery. The world's oldest known sweet chestnut tree is on Sicily's Mount Etna, and it is between 2,000 and 4,000 years old; Britain's oldest sweet chestnut trees date back only to Tudor times. There are scattered written records, the earliest placing a sweet chestnut in Wales in AD 1113.

Sweet chestnut comes from the same long-lived family as oaks and beeches, and its potential lifespan is 700 years. It grows across Britain, particularly in southern England, in light, non-chalky soils and is common in copses and woods. It is coppiced for poles, notably in Kent's woodlands, where the tree is widespread and a major source of the hop poles used by brewers. Lighter than oak wood, which it resembles, its timber is valued for carpentry, furniture and joinery. Its tannin scent is fruitier than oak. There is a lot of it: a mature tree might be as much as 35 metres tall; it could also have a circumference as large as six metres. The grain of sweet chestnut wood is straight, but older trees have a spiral grain.

Its smooth, purple-grey bark develops upward-spiralling fissures and ridges as the tree ages. It has purple-brown twigs with very small, red-brown, oval buds. Sweet chestnut is a deciduous tree, and it has very large leaves, 5-9 centimetres wide and 16-28 centimetres long, sometimes even larger, with pointed tips and sharply serrated edges. The leaves' glossy upper surfaces feature very prominent veins.

By late June or July, sweet chestnut trees are in bloom, with mostly male yellow catkins, the females tending to grow at the bases of the males. These catkins are long and vertical and give off a strong smell that some people characterise as like fried mushrooms but others identify as semen. Whichever it is, it attracts bees and other insects, which assist the wind in dispersing the pollen released by the catkins, in exchange for the flowers' plentiful supply of nectar.

On trees of 25 years or more, pollination enables the female flowers to transform into spiky green seedcases or husks, which enclose the well-known shiny, red-brown nuts. These appear in August, the husks opening – or falling from the trees and splitting open – to release the chestnuts during October. While we're contemplating roasting them, red squirrels, mice, birds and micromoths and possibly deer and wild boar are feeding on them raw. The bark of sweet chestnut saplings can be used for cordage or folded basketry. The inner bark from fallen trees starting to decay can be used as a fire tinder.

8

METAL ON
WOOD

Axes such as this were made and used in Britain between 1400–1275 BCE.

The British Bronze Age is nebulous. We can only see a tantalising glimpse of it, like the crown of a mighty oak tree emerging from the mist in a river valley. Archaeologists aren't even certain when it began. Some time around 2500 BCE, it evolved from the Neolithic. Copper tools began to be made, and these in turn would become bronze, with the discovery that the addition of tin made them harder and more durable.

The use of metal tools supercharged British agriculture. Societal organisation, cooperation and communication brought new advances in science and engineering. Compared to the Neolithic, it would have seemed a fast-paced, 'new age' of enlightenment. These were the people who would invent the wheel and complete Stonehenge and, while I have pointed the finger of blame at Neolithic farmers wielding polished flint axes for the reduction of our forests, the arrival of bronze axes would have been akin to replacing a steel axe with a chainsaw. This was certainly a time of increased felling. It has been estimated that during this period, Britain's tree coverage was reduced by 50 per cent, and that woodland management practices such as coppicing and pollarding began in earnest to supply the forest materials required by the farming community.

Coppicing is the cutting of a mature tree close to ground level and the existing root structure. It enables vigorous new shoots to grow, resulting in a new, multi-stemmed tree that can itself be harvested – whole or in part – when of sufficient dimensions. Once begun, the process can be repeated and, in some cases, trees managed in this way are believed to have survived for around 2,000 years.

Coppicing works especially well with hazel, ash, alder, lime and hornbeam, which historically became Britain's principal coppice woods. Coppicing provided ideal materials for the community: ash for its structural strength and for tools; alder for its durability, particularly in wet conditions; lime for its bark and light, strong wood, perfect for shields; hazel for its flexibility and versatility, ideally suited to the wattle walls of the round houses of the time; and hornbeam for its strength and for the intense heat of its charcoal, just right for metalworking. Amongst the coppice, tall standard trees could be grown to provide future coppice trees, along with heavier species such as oak to provide timber for larger projects and boat building. Oak was a favoured wood for dugout canoes because of its ability to cope with waterlogging.

Pollarding is a similar process to coppicing, but the cutting is done above ground, at a level higher than the browsing range of livestock. This allows the woodland to be used as a pasture as well, and it can provide abundant materials for basketry and firewood. Trees favoured for pollarding were willow, beech, hornbeam, lime and oak.

We should not imagine that Bronze Age life was all roses and barley. The impact of bronze blades on trees was matched by their repercussions on human society. This period would redefine concepts of land ownership, witness fluctuating tribal boundaries and land disputes, and productive fields cut out of the forest would need to be defended. While arrow points were still being made from easily available flint, they were to become barbed, a feature designed to convey menace to a human adversary. Indeed, Ötzi, the most famous ambassador from the Bronze Age, who was discovered melting out of a glacier in the Ötztal Alps in 1991, had been involved in hand-to-hand combat shortly before his death. He likely bled to death from a wound inflicted by just such an arrowhead, which today is still embedded in his left shoulder, where it remained when the arrow shaft was pulled out.

Most tantalising of all Bronze Age finds are the astonishing treasures unearthed from tumuli, the burial mounds that dot our landscape. There are stunning everyday items, like bronze shears, and also beautiful, refined items, like jewellery made from silver and gold. This was a time of bling, when warriors carried elegant swords and spears of polished bronze that glinted in the sunlight on the horizon. Even a favoured horse might be adorned with gold. But these people were also pious, generous in their offerings to their gods.

To my mind, this was the last hurrah of British prehistory. There was a continuing sense of spiritual connection to the land, a blossoming in arts and crafts, and probably new stories and songs. It was very much a time of curiosity and ambition; the wheel of progress had truly picked up speed. Perhaps even at the time of Must Farm's demise there was already a sense of anticipation; whispers brought on the lips of traders spoke of the existence of a fabled, new, different-coloured metal that far exceeded even the strength of bronze. Had people already started looking towards the horizon of future inventions, as we do today?

In 1922, Howard Carter excavated the tomb of an Egyptian pharaoh, Tutankhamun. Aged just 19, Tutankhamun had died in 1323 BC and was entombed with more than 5,000 grave goods for his journey into the afterlife. Discovered virtually intact, his tomb is still considered a wonder of the Bronze Age and one of the greatest archaeological finds of all time. Found resting on the departed pharaoh's right thigh was a dagger. It had an ornate hilt with a crystal pommel fitted with gold pins, a grip of gold, inlayed with coloured stones and glass, and a beautifully embossed gold sheath.

More astonishing than any of the decoration, though, was the blade. When drawn from the sheath, it was not of gold or of bronze, but of iron. Even more amazing, despite having been entombed for 3,000 years, it was still sharp, showing only a few minor surface spots of rust. This, clearly, was a special blade. Modern analysis has since established that the iron blade also contains 11 per cent nickel and 0.6 per cent cobalt, which explains its stainless nature and reveals that it was indeed special, made from an iron meteorite. It is not a lone iron object found within the Bronze Age, but it is perhaps the most impressive. There exist a scattering of such finds that demonstrate that Bronze Age metalworkers were already curious, actively seeking new ores and potential new metals.

In 2018, I was privileged to see at first hand the stored ancient timbers of the Seahenge. As their protective wrappings were pulled back I was astonished, for, as fresh as the day they were felled, those ancient timbers still preserve deep, confident axe cuts that were obviously made with sharp axe blades. Having grown up in an age of steel tools, I was forced to confront my prejudice against tools of bronze. The effectiveness of those ancient axes was undeniable. So why was it that the arrival of iron tools would prove to be such a technological revolution?

Ironworking is believed to have originated in the Middle East. The oldest bloomery yet found is at Tell Hammeh, Jordan, and dates to around 930 BC. In Britain, ironworking began some time in the eighth century BC. Iron is considerably more difficult to produce from ore than bronze but, once made, it offers many advantages compared to bronze. As a raw material, it is more abundant than copper and, particularly, the rare tin needed to make bronze. Iron is stronger, more malleable and easier to work with than bronze. An incredibly versatile material, iron can be immediately manufactured to meet the user's demands, and quickly modified following experimentation, and it is often repairable.

As ironworking skills improved, smiths learned to fashion it into ornate shapes for many purposes. With the addition of the correct amount of carbon, it could be used to create enduring sharp edges. By controlling the temperature of its forging, it could be softened, hardened, made more durable or even springy. In time, blacksmiths would master pattern-welding, allowing billets of steel exhibiting different desired qualities to be combined within a single tool. This new metal was amazing beyond imagination. Tools of iron would become ubiquitous, changing societies' outlook on metal and the world around them for ever. It would usher in an age of pragmatic design and provide the tools for an even higher degree of craftsmanship. Bronze would not be forgotten, but it could now shine in roles better suited to its qualities, including the production of beautifully ornate, polished bronze mirrors, that would travel with their high-status owners in Iron Age burials.

Mastery of iron issued a greatly improved and expanded toolkit. In addition to sickles, scythes, knives and axes would come draw knives, froes, planes, gouges, chisels, gimlets, augers, drills and saws. These tools would combine the sharpness of flint tools with the toughness of bronze, without any brittleness.

A new fastening – the nail – would also make its debut, a humble revolution, but one through which the construction of wooden buildings and ships could be achieved with much-improved speed and efficiency. Sharp tools did not need to be sharpened so frequently, and the hardest woods could now be easily worked. Finer woodwork could be achieved more quickly and to a more refined finish. Saws would permit a more efficient use of timber too; instead of hollowing out logs, boats could instead be lightly built of planks. Although this was not a process new to the Iron Age, planks could now be sawn rather than shaved down. With iron tools, the carpenter could work more economically and maximise the properties of the wood he worked, sawing planks strong enough for his purpose but without unnecessary bulk. These could then be bent to more graceful and more seaworthy designs.

Iron Age ingenuity would see the first powered machines: the pole lathe and the potter's wheel. Both were revolutionary new inventions. With each evolution of metal tools, new possibilities for our use of wood emerged and our carpentry skills and ambitions grew, placing ever-increasing demands on our forests to supply materials. Human ambition soared. Eyes lifted from the land to the possibility of new lands across the horizon. In only a short while, we would be shipping in new types of timber from overseas to take advantage of their exotic characteristics. Any spiritual kinship we once felt for the forest seems now to drown under the wood chippings and shavings of the new workshops. Trees were simply there to be taken advantage of, grown for purpose, harvested and used. This was the true beginning of the modern age.

Today the Iron Age feels like the distant past but, arguably, it is still within living memory. At the time that iron first appeared in Britain, the tiny seed of a yew tree germinated in Scotland. Successfully contending with the dangers any seedling must face, it grew successfully into a tree and, by AD 43, when Julius Caesar contemptuously laid his sandal on the coast of Kent, it was already what we would classify today as ancient. Astonishingly, that tree is still alive and growing even now. The Fortingall Yew in Perthshire is believed to be Britain's oldest tree, at around 2,000–5,000 years old. What an incredible period of history it has lived through.

To me, it is a representative of the many mighty trees that once were, but which we felled in our enthusiasm for new tools and technologies. Ancient trees have a very special place in the forest. Over the course of their incredibly long lives, they become unique habitats; a repository of specialist wildlife essential to the biodiversity and health of a natural forest. I like to think of them as a source, a spring of forest life, essential hubs in the complex web of the forest both above and below ground.

For our hunter-gatherers, such trees would have stood out in the forest, particularly in winter when their dark, evergreen crowns would have provided impromptu shelters, with deadwood for a fire and a ring of fallen branches that could easily be arranged to keep out dangerous game. To stand under an ancient yew tree today is to stand in a sanctuary, a place of quietude and contemplation. It is believed that Neolithic farmers venerated ancient yews as sacred places, and strange ancient country rites associated with the yew survived into recent history. Were they once places of worship? Today, if you can, stand beneath an ancient yew tree when the sunlight is slanting through the tall stem and arch-bowed branches. You will feel a cloistered atmosphere.

Was this the inspiration for the architects of our great cathedrals? To me, such trees are temples of nature, places of worship, where I can feel a deeper connection with the natural world, with past times and where I can consider the possibilities of the future.

Sycamore

While they may no longer be the popular childhood activities they once were, a British autumn still regularly offers a reliable pair of natural delights. September brings the fruit of the horse chestnut (*Aesculus hippocastanum*), plunging groundward in spiked cases that split on impact to reveal the conker inside. Then, each October, the sycamore tree (*Acer pseudoplatanus*) releases its seeds to flutter softly to the ground. The first sight of these, for a child, is an arresting one: a pair of brown wings, spinning very gently as they fall, sometimes caught on the breeze and shunted a little further from its parent, its rotating blades picking up a little speed. In the botanical world, they're called samara; in days of yore, their colloquial names included 'keys' and 'spinning jennies'; during the twentieth century, they became known as 'helicopters'.

The seed of the sycamore tree is actually a double samara – two single-seeded 'wings', joined together at an angle to form a symmetrical V-shape. Each wing is 20–40 millimetres in length and rounded at its end, its width narrowing towards the join, where the seeds are. Their falling spin is not unlike the whirling blades of a helicopter, if considerably gentler.

The angled wings slow their drop from the branches above and aid their dispersal: shifted off course by the mildest breeze, they'll not all be falling to gather in an unhappy heap on the same spot. As they float down, the air pushes up against the wings in a mild dispute with gravity, and nudges the central join sideways a touch, forcing the wings to spin, slowing the fall. A stronger wind will obviously affect direction and increase the distance travelled, increasing the seeds' chances of finding somewhere to generate new trees.

For generations of schoolchildren, this presented an opportunity to grab the spinning wings and chuck them back up in the air, perhaps attempting to race them – whose will hit the ground first, or last? (It wasn't a long game.) Now, the sycamore fruit's aerodynamics have been absorbed into the National Curriculum's Key Stage 1 for primary school pupils aged five to seven, something which, somehow, simultaneously deadens the joy while also rightly bringing a little of nature's excitement into the classroom.

In Britain, sycamore shares this unusual method of dispersing its seeds with ash and a couple of maple species. Native to large parts of Europe, the sycamore was a late arrival in the British Isles, probably introduced here in Tudor times, or just possibly by the Romans. The species may have become naturalised in the 1600s.

It can live for up to 400 years and may reach 35 metres in height. Its smooth, young bark is a pinkish-grey that develops fissures and becomes rough as it ages, and its hairless twigs are pinkish-brown. Five-lobed palmate leaves measuring up to 16 centimetres grow on red leaf stalks. After pollination, its yellow-green flowers blossom from April to June to hang in spiked clusters called racemes, before developing into the winged seeds that ripen on the branch in September and October.

Sycamore yields a hard, strong timber, which is perfect for carving. The fine-grained, pale-cream wood is used to make wooden spoons, ladles and other kitchen utensils, because it will not contaminate food. In Britain's uplands, abandoned cottages and homesteads often have an old sycamore growing nearby.

The smoke from the bark has a stupefying effect on bees and was used in monasteries to smoke truculent beehives. Sycamore wood can also be used for friction fire-lighting, and the leaves for wrapping food or lining ground ovens.

Walnut

The binomial nomenclature for walnut is *Juglans regia* – literally, 'royal nut of Jupiter'. The ancient Romans knew it as *Jovis glans* – 'glans of Jupiter'. Roman mythology held that, when the god Jupiter was on Earth, he fed on the fruit of the walnut tree, probably because his slightly older forerunner in ancient Greek mythology, Zeus, actually had the walnut dedicated to him.

Once again, this is a tree that is not native to these isles but which might well have been here since Roman times. It was not, however, reported as naturalised until 1836. A deciduous, broadleaved tree, it now grows wild in Britain's lowlands, secondary woodlands, on riverbanks and sometimes at roadsides. It is also planted in parks and gardens, grown for ornamentation, for its nuts and, principally, for its timber, which is a fine wood with a wavy grain and is used for high-end furniture.

Walnut trees prefer well-drained loam, fertile and alkaline soil, and will grow to 35 metres tall at their peak, despite their short trunks. They have broad crowns but will grow narrower ones if in limited woodland space.

From mid-May, yellow-green male catkins with as many as 20 deep-purple anthers droop from curving green twigs. If these thickset twigs are snapped or cut open, they reveal a sponge-like inner pith. After they have released their pollen in June, the catkins drop from the tree. Once the females are pollinated, they develop into fleshy green seedcases, each one enclosing a wrinkled brown walnut. These seedcases start to appear in August and are ripe by autumn, when mice and squirrels will start to forage for them.

A beautiful wood to carve, walnut has a dark colour and responds to sharp tools. Holding the finest detail, it was often chosen for carving love spoons. It is also resistant to splitting. Both of these qualities fitted it to the production of musket and rifle stocks. During the Napoleonic period we nurtured plantations of walnut to ensure a supply of stock wood. The dark-brown colour inspired the soldiers' name for their famous musket of the time, the Brown Bess.

Wayfaring Tree

In 1597, the herbalist John Gerard published *The Herball, or General Historie of Plantes*, later known as *Gerard's Herbal*. Gerard describes *Viburnum lantana*, calling it 'the Wayfaring tree' and noting its presence 'in the chalky grounds of Kent about Cobham, Southfleet, and Gravesend, and in all the tract to Canterbury':

> The Wayfaring man's tree grows up to the height of an hedge tree, of a mean bigness: the trunk or body thereof is covered with a russet bark: the branches are long, tough, and easy to be bowed, and hard to be broken, as are those of the Willow, covered with a soft whitish bark, whereon are broad leaves thick and rough, slightly indented about the edges, of a white colour, and somewhat hairy whilst they be fresh and green; but when they begin to wither and fall away, they are reddish, and set together by couples one opposite to another. The flowers are white, and grow in clusters: after which come clusters of fruit of the bigness of a pea, somewhat flat on both sides, at the first green, after red, and black when they be ripe: the root disperseth itself far abroad under the upper crust of the earth.

Gerard's work was criticised for errors before it was even printed and was revised soon after his death. Now we meet a passing 'countryman of Essex', who identifies the plant as 'Cotton tree', 'by reason of the softness of the leaves'. No edition mentions that the tree's berries are poisonous, nor that its bendy stalks were used to tie hay bales.

Its old country name was 'hoarwithy', indicating its use as a binding material, and in prehistory its straight, rigid wood was chosen for arrows, as we learned from Ötzi. Long saplings also straighten well over fire and may well have been a good choice for spear shafts.

Whitebeam

Whitebeam (*Sorbus aria*) is native to southern England and is now also planted in the north of the country. Though not often seen growing wild now, it remains common as an ornamental tree in parks and gardens. It grows best in chalk and limestone.

Its young twigs are hairy but lose their hairs and become smooth when older. The green leaf buds are pointed and have hairy edges. The irregularly serrated leaves, when they emerge, are also noticeably hairy, a thick white felt coating their undersides, but a shiny dark green on the upper surface, which will fade to russet before they drop in autumn.

Clusters of flowers blossom in May and await pollination by insects. Each five-petalled flower is hermaphrodite, part-male and part-female. By September, the flowers have grown into scarlet fruit, which ripens through autumn. The red fruits look like large berries, but they are in fact pomes. They are too big for many birds: blackcaps, robins and starling cannot manage them; blue tits, chaffinches and green tits peck at them. They're enjoyed by all types of thrush, however, and are also popular with crows and woodpigeons.

Perhaps because so many bird species are unable to eat them, lots of whitebeam berries remain on the tree for the whole winter, blackening as they start to decompose. Oddly, the raw fruits become edible to humans when almost rotten; before that, they need bletting. In north-west England, they are known as chess apples.

Whitebeam's hard, white timber has a fine grain and is suitable for woodturning and fine joinery. It was once used for machine cogs until iron superseded it.

9

WEAPONS

The hooked end of a *pitjantjatjara miru* or spear thrower, lashed together with emu leg sinew. Spear throwers pre-date the bow, their effectiveness well reflected in their continued use in Australia today.

Strange as it may seem, the oldest known wooden implement worked by a human was found at Clacton-on-Sea in 1911. It is a piece of yew wood, 39 millimetres in diameter and slightly less than 39 centimetres long. Broken at its thick end, it has been carefully tapered to a fine, sharp point. It has been dated to 420,000 years and is very likely the tip of a broken spear or javelin.

Ten other similar and more complete spears were found in Schöningen, Germany, in the 1990s. These spears, which have been dated to between 337,000 to 300,000 years old, indicate that their Neandertal makers well understood the properties of wood. Made from slow-growing spruce, with one of pine, they were carefully worked, debarked and tapered at both ends. Their tips were fashioned at the base end of the wood where the wood is densest and hardest, offsetting the point deliberately so that the soft-core pith was not at the tip. They vary in length from 1.84 metres to 2.53 metres with maximum diameters from 29 to 47 millimetres.

Additionally, some of these spears are tapered like a modern javelin, with their thickest diameter in the first third of the spear. One of these spears was discovered deeply embedded in the pelvis bone of a wild horse, a clear indication of their purpose. It seems likely that those ancient hunters were ambushing wild horses in boggy ground. Recent experiments by archaeologists from UCL, have shown that these projectiles could be thrown with sufficient force to kill at 20 metres. Crude as they may seem, they are fascinating. Were I to find myself in remote country where there were large, dangerous animals, they'd be precisely the design of spear I might improvise quickly for my protection, a weapon that can be used in a thrusting manner to fend off a predator and with sufficient aerodynamic properties to be cast accurately.

Perhaps the Clacton spear had served just such a purpose; there were plenty of dangers posed by predators at that time. Did the maker also use that spear to hunt straight-tusked elephants? They were twice as large as the elephants walking today, weighing up to ten tonnes. That the Clacton spear is made of yew is also interesting. In British folklore, yew has long been known as the tree of death and historically has proven to be one of our most important trees.

Javelins would receive a technological boost from the invention of the spear thrower. The oldest spear-thrower yet found, from Combe Saunière in the Dordogne, has been dated to 19,000–17,000 BP, but it is probable that the weapon is much older than that. Best estimates suggest a date of around 30,000 years ago, although a 42,000-year-old skeleton found near Lake Mungo in south-eastern Australia shows elbow wear consistent with that presented by people who have used a spear-thrower all of their lives.

Most spear-throwers are made from wood, hence their rarity, but archaeologists in France have found Late Palaeolithic spear-throwers preserved in rock-overhang shelters and in caves. They survive because they are not made from wood but from more enduring reindeer antler. Beautifully carved, they are astonishing works of art that enable us to step into the minds of the hunters who used them. While a few are considered complete, most were intended to be fittings on a wooden haft. They are truly extraordinary objects, often fashioned as complex three-dimensional figurines depicting all manner of wildlife; grouse, wild horses, bison, fish, eel, ibex, felines and mammoths. Some of these figurines are stylised, while others are incredibly anatomically accurate and well proportioned. Three are carved to depict exactly the same design of an ibex giving birth. Was this a cultural identifier? Did the depiction represent a folklore story? We can only guess. When I look at these wondrous works of art, I think of the engraved stocks of big-game rifles, on which are depicted scenes celebrating a hunter's reverence for the prey. Surely here we have depictions of the many varied animals those ancient hunters stalked. But more than that, these sculptures were carved with a confidence suggestive of the hunter's mastery of the animals represented. We are fortunate indeed that these items survive, but how many others were carved from wood that have not survived?

Wooden spear-throwers have, though, survived in the Western United States, particularly where preserved by an arid climate. These spear-throwers, often referred to as atlatls, can be intricate, fitted with weight stones to enhance their performance.

My own experience using spear-throwers in northern Australia has taught me how great an improvement they are over a hand spear or javelin. The advantages are many and subtle. In the first instance the extra leverage results in a faster projectile that enables the hunter to use lighter, longer-range spears, which can be fitted with specialised projectile points to suit different applications. The spear-thrower also provides greater reach with the spear while promoting far improved accuracy. But perhaps most importantly of all, and rarely mentioned, it enables the hunter to strike with speed faster than the game's reflex to flee.

A hunter equipped with a spear-thrower can carry several projectiles, is able to travel light and fast over long distance and is more able to hunt and travel alone. In my mind, I see the confident, upright posture of my Indigenous tutors and apply it to our Late Palaeolithic hunters. Although spear-throwers can comprise many parts and be complicated in design, they are fundamentally easy to make. Unlike the bow, they do not need a high-quality string, which is not always easy to obtain, and they are more resistant to extremes of weather and climate. Accordingly, they achieved global popularity: they were used by Inuits spearing seals from kayaks in the High Arctic; in the equatorial rainforests of Brazil; in North America to hunt Palaeolithic megafauna; and they are still in regular and diverse use in Australia today.

It may be virtually forgotten now, but I believe that the invention of the spear-thrower was a technological revolution of importance approaching that of the mastery of fire. The spear-thrower made it possible to source meat and fish with far greater certainty. Besides an ambush, communities could now hunt on a more opportunistic ad hoc basis with a higher degree of certainty. This in its own way would facilitate a more mobile lifestyle, following migrating herds and exploring the new landscapes of the post-glacial world. In so many ways the spear-thrower was the perfect weapon for the open lands of Europe immediately following the retreat of the glaciers a time of cold landscapes and a flesh-heavy diet.

Although no spear-throwers have yet been found in Britain, I find it hard to imagine that they were not widely used here. There are many projectile points in our archaeological record which share features with projectiles known elsewhere to have been used with spear-throwers, particularly barbed antler points, around 250 of which have been found in England. Of those, 227 were found at one archaeological site, Starr Carr in North Yorkshire, where the conditions for their preservation were favourable. These points could of course have been thrown by hand but they are exponentially more effective with a spear-thrower. At some point, the remains of a spear showing the characteristic dimple in the tail end that articulates with the peg of a spear-thrower will be found, if not a spear-thrower itself.

My own investigation into spear-throwers has revealed how incredibly diverse they are in shape and form. Perhaps those early British hunters used spear-throwers like those found in France. But what of the spears or darts they propelled? From French finds, we know that the points of the spears varied greatly: sometimes a flint blade was glued and bound to the point, while others were more complex, with pointed fore-shafts of bone and antler, scarf jointed to the main shaft. Some had the fore-shafts fitted with multiple flint blades glued into slits carved into the fore-shaft's edges, while others were made of antler carved with a serrated edge. What is missing is the spear shaft that connected to the spear-thrower.

Possible clues to what these might have looked like have been found in the mountains of the Yukon. There, montane ice patches have been melting in the face of global warming, and hidden in these ancient snowbanks are archaeological treasures. In 1997, a small, piece of wood was found. On investigation, it was recognised as the fore-shaft for a spear-thrower dart. It was 4,000 years old. Many fragments of ancient darts have been found since then.

In 2018, a complete dart, two metres long, was recovered; it had last been seen by a hunter as it buried itself in the snow 6,000 years before. The state of preservation was incredible: apart from having been distorted, it was otherwise as fresh as though it had been made yesterday. Analysis of the object has revealed its sophistication. It was made from three lengths of birchwood scarf jointed together and bound with sinews. While many of the darts found are made from saplings of a suitable size, mostly from willow or birch, this shaft was made from a birch tree that had been split into quarters. These had then been carved down to make the individual segments of the shaft. The point was a knapped flinthead, fitted into a slot with raw spruce gum and bound with sinew. At its tail end, the dart was fletched with eagle feathers for stability. The vanes had been cut down from one side of each feather and the remaining vanes attached with four bindings of sinew, some of which passed through holes made in the feather quill. It was a work of art.

Could this dart and the many others found in the American Northwest represent the equipment our Palaeolithic hunters used? I think it is quite possible. The landscape in which the Yukon darts were cast was very similar to post-glacial Britain. The tree species represented in the Yukon finds, birch, willow and spruce, were virtually identical, and both hunting communities were seeking reindeer amongst other game. The full range of fore-shafts found in North American ice patches have a variety of points still attached. These are very similar, virtually interchangeable, with those found in Europe.

I can imagine the hunter on the day that dart was lost, making a concerted search for it, raking through the snow. But even as his search was being made, the next great revolution in weaponry had already been in use in Europe for 6,000 years, if not longer: the bow and arrow. A more compact weapon with greater range and better accuracy that would soon make the spear-thrower obsolete everywhere, except for Australia.

BOWS AND ARROWS

The longbow is a stick, nine-tenths broken, shaved back sufficiently to allow it to bend but not so far that its spirited willingness to straighten again is destroyed. It is at its most vulnerable at that moment when it is pulled back to full draw, the extent of its intended arrow. Because of this, in contrast to modern composite bows, it will only be held at full draw for a heartbeat's duration before the archer's strained fingers smoothly straighten, allowing a crisp, swift release of the arrow. How often English history turned on an archer's heart-beat: Hastings, Crécy, Poitiers, Shrewsbury, Agincourt, Towton …

Aiming and release by an experienced longbow archer were virtually instantaneous, a skill built upon a lifetime's practice at the butts, the official archery training ground once found in every village, and more significantly by roving – shooting at often tiny targets in the countryside in a spirit of friendly rivalry. At the Battle of Crécy in 1346, massed longbow archers dramatically outshot the Genoese crossbowmen employed by the French army, cementing the presence of military archers in English armies for the next 200 years.

On 21 July 1403, archers stood once again massed before two opposing armies, but this time facing each other on the outskirts of Shrewsbury. This day would witness the terrible effect of archers in a battle between fellow Englishmen on English soil. The rebel army, the larger of the two armies, was led by Henry Percy, the first Earl of Northumberland better known as 'Hotspur' and Thomas Percy, the first Earl of Worcester. They were well supported by a retinue of renowned Cheshire archers, some of whom may well have seen service in Ireland and France. The smaller force comprised the army of Henry IV and, to his left, the army of Henry of Monmouth, the Prince of Wales, not yet 16 years of age.

Two hours before dusk, following disappointing and inconclusive negotiations, Henry IV decided to launch an immediate attack. Thus, the Battle of Shrewsbury began with massed arrow volleys from both sides causing heavy casualties, before the men at arms closed for combat. During the battle, the Prince of Wales received an arrow in the left side of his face but, fearing his withdrawal from battle would cause his men to lose heart, he disregarded his injury and fought on. We are also told that during the melee Hotspur pushed up his visor, perhaps to improve his vision or to issue commands, and in so doing exposed his face to a serious arrow injury. The Battle of Shrewsbury was won by King Henry, and once again the potency of the bow in war was reinforced at terrible human cost.

The Prince of Wales's subsequent recovery was fortuitous indeed for England. As King Henry V, he would prove to be one of our greatest kings, chivalrous, courteous and a brilliant battlefield commander. In 1415, he would stand on the field of battle, once again facing impossible odds but this time in France near to the village of Azincourt. I wonder has any other king in history truly understood the potency of the bow in the way Henry V did? His army was 80 per cent composed of archers. Agincourt would be a great victory for Henry V, England and his bowmen. It is hard today to understand how great an achievement this was judged to be. On the king's triumphant return to London, he was greeted by the greatest pageant of the age. The city's water conduits, fashioned from whole elm trees bored through, did not flow with water that day but were instead filled with wine.

In 1982, like many of my age, I sat and watched transfixed as the *Mary Rose*, a Tudor warship that sank in the Solent while engaged in battle in July 1545, was lifted from the seabed on a giant yellow gantry. Today you can visit the *Mary Rose*; she is exhibited in Portsmouth's Historic Shipyard. As if travelling through time, you can gaze upon the preserved ship and wonder at the astonishing artefacts that were recovered from the ship. One of the benefits of the excavation of the *Mary Rose* would be the light it would shine on late medieval military archery.

From the ship's inventory, we know that she was equipped with 250 longbows and just short of 5,000 arrows when she set sail. From the wreck site, 172 longbows were recovered, 2,303 arrows and 7,834 arrow fragments. Most of the bows were found in an astonishing state of preservation: their horn nocks had rotted away, but the yew staves themselves had survived in incredible condition. The best-preserved bows were those recovered in elm storage boxes, 50 bows to a box. They look as though they were made yesterday. Those bows found strung for action on deck were more exposed to the sea and have not fared so well.

With such an incredible sample of medieval bows, our understanding of the medieval war bow, the men who drew them and the allied trades has been forever changed. Until this quantity of military bows was recovered, the few that stood elsewhere in cabinets had been largely overlooked. Many archery authors from outside of Britain, when writing about the British longbow, had based their assessments on the beautiful slender target bows of later centuries, perhaps also assuming that these bows were more evolved than the rather crude-looking medieval war bows pictured in medieval illustrations.

But how wrong they were. Studies of the *Mary Rose* bows have shown that, although they are unadorned, they are superbly crafted. One of their mysteries is that they are reflexed, that is bent very slightly in the wrong direction in the handle. No one has yet determined whether this is a by-product of five centuries underwater or, as seems more likely, a deliberate feature of their manufacture. Assuming that the bows found aboard the *Mary Rose* are typical war bows of their age, their draw weights are estimated to have ranged between 65 and 160 lb, averaging at 110 lb. I have long wondered whether the standard required for military bow manufacture of the age was a 100 lb minimum. Did a wise bowyer perhaps opt for producing a heavier draw weight to ensure his bows were accepted and to account for any drop-off in power while in use or while stored in a damp building?

What is immediately apparent from the *Mary Rose* is the scale of the industry required to maintain archery supplies for warfare.

Consider that there were 250 bows and 5,000 arrows aboard just one fighting ship, to say nothing of the rest of the fleet, castles, fortresses, and garrisons across the realm. All of the bows aboard the *Mary Rose* were made from yew. This is Britain's superlative wood for bow-making, but it is not our only suitable wood. Henry VIII had required that for every yew bow made, a bowyer should fashion at least two bows of wych elm or other woods of less value, under penalty of imprisonment for eight days. This suggests that suitable yew was in short supply for military needs. What happened to the lesser bows? Were they used in town garrisons and fortifications?

Beyond the needs of the military, Henry VIII, following on from the decrees of his predecessors, required all his male subjects under the age of 60 who were not decrepit or maimed or having other lawful impediment to practise in the use of a longbow. Failure to do so could incur a fine of 12 pence per month; the clergy and judges were excepted. Every town was ordered to provide butts to practice at; the name 'butts' is still visible in village street names today. Parents had to provide every boy from 7 to 17 years of age with a bow and two arrows. Once the boy reached 17, he was required to furnish himself with a bow and four arrows. Foreigners were not allowed to shoot a longbow without a licence, while the crossbow was proscribed unless you earned more than 200 marks per year. Everybody was shooting arrows. The supply of materials for all of this archery must also have had a significant impact on the management of our woods and coppices.

In many ways, yew is a fragile, brittle wood compared to other, more stringy, grained bow woods such as wych elm and ash. The fact that all the *Mary Rose* bows were of yew is clearly not accidental and speaks volumes regarding its performance and the regard with which it was held.

There are several advantages to using yew. Firstly there is natural lamination within the wood. The cream-coloured outer sapwood has good resistance to tension while the heartwood resists compression. By making a bow with some sapwood intact, the tension loading of the heartwood moves more towards the heart of the bow where it is less vulnerable. But regard must also be paid when working yew to not overload the bow by leaving the stave too thick at any point; to do so will result in a chrysal on the bow's belly, literally a crease where the grain collapses inwards under the compressive load. This is not to say that yew bows must always be made with the sapwood intact; bows can be successfully made of yew with only heartwood. Just such a bow was found in Somerset, the Mere Heath Bow, dated to 2690 BC.

The often understated advantage of yew is the lightness of the wood. Compared to other bow woods, they are very light in weight. This allows the limbs to spring back faster than heavier more sluggish woods. One has only to walk into the low boughs of a yew tree at night to feel both its spring and resilience. The yew is a thoroughbred tree, with the perfect combination of virtues: light, fast limbs, the advantage of the natural lamination and a potent springy demeanour.

Against its use as a war bow were two factors. First, it can be a difficult wood to work. Its pedigree nature means that it tends to exaggerate everything the bowyer does, rewarding good workmanship but mercilessly punishing any momentary lapse in concentration. To overcome this, bowyers were carefully trained and forbidden from working after dark.

The second Achilles heel in the yew war bow was its very slow rate of growth. Yew was being cut and used faster than it could be regrown. Yew suitable for bow-making is uncommon, and it needs to be of the correct size, free of knots and straight grained. Even today, the majority of yew bows are made from two short sections of yew fishtailed at the handle, because long, clean-grained yew that would make a single-stave 'self' bow remains a rarity. To overcome this shortage of supply, a tax was eventually levied on imported goods, which required a duty to be paid in yew staves suitable for bow production. Perhaps this dependence on foreign-sourced yew spurred the belief that our domestic yew was inferior, but this is clearly false. My own bow of native English yew shoots as well as foreign yew.

From the *Mary Rose* we have also learned that the arrows were stored below deck, bound in sheaves of 24 arrows, and that these were decanted into arrow bags fitted with pierced leather spacers for use on deck. One must envisage the process of loading these bags and the passing of them up companionways to the waiting hands of skilled archers. But imagine also the massive demands such equipment placed upon the logistical supply chain of the country and the cost of equipping a late medieval army with archery equipment. Bows aside, the arrow itself was a complicated and costly projectile: each arrow was comprised of six component parts.

1. The stele or shaft

Most of the shafts aboard the *Mary Rose* were made from poplar wood, a quick-growing hardwood that is light and strong, and can be readily riven into boards for arrow production. In many ways, it was the perfect choice for the mass production of arrows. Birch and alder shafts were also found, both also fast-growing woods of the coppice with a long heritage in arrow-making, being suitably stiff for a strong bow, straight-grained and easily worked. Other woods represented in the *Mary Rose* finds include hornbeam, willow, elder, walnut and ash.

While modern arrows are milled by machine into beautifully parallel sided shafts, the arrows of the *Mary Rose* showed greater variety of design. Most were tapered from the point to the tail, what is called a little-breasted or a bobtail arrow. The next most numerous in the finds were parallel-sided arrows, the easiest to produce. But there were also barrelled arrows thicker in or around the midshaft, tapering to point and nock, considered by some to be the best arrow shaft design of all. Then there were some big-breasted arrows that seem to defy logic, tapering from nock to point.

These arrows were either 76 centimetres or 71 centimetres long and stored accordingly. Does this mean that these archers were all drawing to the chest to pull the full length? Perhaps, or perhaps they simply drew to their own individual draw length, in which case, the majority would be safely accommodated by the longer arrows. Thinking pragmatically, if I had to equip an army with arrows, I would equip each archer with slightly longer arrows on the basis that some will break in practice or in use. The most common breakage results from directly striking a solid object causing the shaft to snap cleanly behind the ferrule of the point. Long arrows could easily be repointed for use. Arrows too short risk being drawn inside the bow where they may act against the grip of the bow, causing it or the arrow to break or possibly backfire.

2. The nock insert

The nock is the slot cut into the tail of an arrow to accept the string. The *Mary Rose* arrows had nocks reinforced with a sliver of horn, glued into position. Set crossways to the grain, this addition would reduce the risk of the nock splitting in use. Throughout history, however, arrows have been made without a horn insert. This simple fitting, while understood, is still perplexing. To make, fix and wait for the glue to cure when dressing millions of arrows would have placed a great strain on arrow production. Clearly this simple fixture was considered important. Was there a now forgotten incident when arrows without inserts had failed? Was its purpose to standardise the nock width for the strings supplied?

3. The glue

Glue was required to assemble the arrow, to affix the nock sliver, the arrow point and the feathers. The glue found on the *Mary Rose* arrows was fish glue containing verdigris, the green oxide of copper. Medieval illustration of battles clearly show the green of verdigris beneath the fletching of English arrows. An astonishingly small detail that perhaps attests to the accuracy of other details seen in these paintings, such as arrows deeply penetrating armour.

4. The arrowhead

Although produced en masse, each arrowhead was forged from iron by the skilled hands of arrow smiths. The arrowhead designs of the age reflect changing trends in armour and the need for mass production. Official patterns for arrowheads were sent out to arrow smiths to standardise production.

5. The fletching

Each feather was fitted with three half-goose or swan-feather vanes. These needed to come all from a left or all from a right wing so they would naturally lean in the same direction. These vanes were both glued and bound in place. It is here that verdigris in the glue most likely had its greatest impact, acting as a preservative against mould or feather mite infestation that could easily render a stored mass of arrows useless.

6. The thread

The fletching of the arrows was completed by binding thread. Here, no expense seems to have been spared, red silk thread being the order of the day.

If you love trees and woodland, I encourage you to visit the *Mary Rose* and gaze upon her bows. Epitomising the height of war bow manufacture, they are stout-limbed bows designed to shoot heavy war arrows, simple sticks behind which we stood as a nation for more than a hundred years. The bow and the yeoman archer have come to symbolise the best of our national virtues: the ordinary citizen, stout and true, willing to stand and face adversity, to fight ceaselessly against tyranny and injustice. Little wonder Robin Hood was an archer.

Despite the archery equipment in use aboard the *Mary Rose* in 1545, the death knell of British archery had already sounded at the Battle of Crécy in 1346, when, alongside his massed archers, Edward III deployed cannon. By the reign of Henry VIII, handheld guns were already capturing the imagination. Britain's long history of archery, a history that stretches back at least 12,000 years, had reached its zenith.

• • •

The oldest complete bow found in Britain is 11,000 years old. It was uncovered at that remarkable archaeological site in Yorkshire, Star Carr. It is carved from a quartered willow sapling. At the time that it was fashioned, the choice of bow woods available was limited to willow, birch and alder; no yew grew at that location at that time. The finished bow is of rat-tail design and is a diminutive 137 centimetres long and approximately two centimetres in diameter at its thickest point. Was it a child's bow, was it a small fishing bow, or was it for shooting the smallest of game? We shall never know, but it is a truly remarkable object.

The bow was probably a recent arrival, a technological revolution that would come to be more important as the vegetation across Britain thickened. With a bow, the hunter could shoot swiftly and discreetly through small gaps in the vegetation, where a spear was more awkward to use. The archer could also carry many arrows easily and quietly, thus increasing the chance of success, and an arrow missing the game might not always spook it in the way a much larger spear does.

One of the defining features of our Mesolithic hunters was their use of tiny flint blades glued into arrow shafts with birch or pine tar, akin to using fragments of razor blade or craft knives. The archaeological record is rich with such finds so, if just a small proportion was used on arrows, archery must have been prolific in the Mesolithic. Occasionally, these points have been found either still attached to an arrow shaft or lying as they came to rest, with the wood now long vanished. From these extraordinary finds it has become apparent that these small blades were often hafted in multiples. These were hunting arrows. While the arrow had to penetrate, the key requirement was to cause massive haemorrhaging to bring the quarry down swiftly.

Recent work by York University has re-evaluated a Star Carr find originally thought to be a canoe paddle, but now believed to be a bow of birch. A reconstruction based on the fragmentary remain suggests a bow with a draw weight in the 40–50 lb region, more than capable of accounting for the deer species at the time. We can now picture the hunters at that time equipped with a sophisticated, specialised toolkit for hunting, carrying with them either bows of birch with birch arrows tipped with flint blades or antler harpoons cunningly hafted in shafts of willow, birch or even alder. Alder is a straight-growing wood with stiffness and resilience well suited to a harpoon shaft. The choice of which weapon would be determined by the game they hoped to encounter.

As time progressed, the increasing temperature would have made more tree species available to our ancestors for the fashioning of bows and arrows, including hazel, wych elm, ash, hornbeam and box. During the Neolithic, as farming replaced hunting and gathering, hunting continued invariably utilising beautifully knapped, leaf-shaped points of flint. Arrows were made from birch, pine, hazel and the viburnum species, guelder rose and wayfaring tree.

Viburnum were popular woods for arrow-making. In 1876, a viburnum arrow tip with a leaf-shaped blade still glued in place was discovered in Fyvie, Aberdeenshire. Ötzi, the Bronze Age man discovered melting out of an Alpine glacier in 1991 was also carrying arrows of viburnum. The main reasons the viburnum species are well suited to arrow-making are that they grow straight shoots of a suitable diameter which are flexible enough to be straightened by heating over a campfire, and that the wood is also stiff when dry. This makes for a fast, lightweight arrow and more accurate shooting but, most importantly for a hunter, prevents the arrow slapping the bow when it is released. This slap results from the way an arrow behaves when loosed from a bow, which is called the archer's paradox.

When a bowstring is released, the sudden instantaneous loading on the arrow shaft causes it to flex; as the arrow travels forwards it straightens, flexing back violently towards the opposing side. In this way the bent arrow oscillates from side to side, the oscillations gradually diminishing during its flight. If the arrow is not sufficiently stiff, it flexes back before it has cleared the bow, causing it to impact the side of the bow, casting the arrow off to the side and making an audible noise. While an archer may compensate for the deflection of the arrow's trajectory, the sound of the slap travels faster than the arrow and can cause any prey equipped with lightning-fast reflexes to break for safety, either spoiling the shot or evading it altogether. To overcome this, the stiffness of the arrow, its spine weight, needs to be matched to the bow. It can also help to offset the bowstring towards the arrow side of the bow, using what are called side nocks. Interestingly, the *Mary Rose* bows show signs of having been side-nocked. This simple feature would likely have ensured a greater parity in performance between the bows and mass-produced arrows in the military arsenal.

By the stress-filled nature of their lives, bows tend not to be long-lived. Eventually, the changing moisture content in the wood may cause a weakness to fail, or perhaps the bow is drawn by an untutored hand and snapped; I once had a reconstruction of a Mere Heath bow broken in this way. There again, the string may snap causing the bow to suddenly straighten like a whiplash breaking its belly. Rarely are these breaks repairable, at which point the bow becomes just another piece of firewood. For this reason, few bows survive from the past.

One bow which did survive, is to my mind the most beautiful prehistoric bow yet found. A hillwalker discovered it in 1990, sticking out of peat in the hills above Moffat, Scotland. A slender, gracile bow of yew, it had broken at full draw, 6,000 years before.

Did the archer hold too long at full draw, was an attempt being made to draw longer than intended for extra power or did a simple unnoticed flaw in its manufacture cause the breakage? Any or all those scenarios could have been the case. I am certain the archer was sickened by the loss, for it was a wonderful bow.

The bow was discarded where it broke, evidence of the hunter's presence on a hillside and in a location that still to this day would favour a hunter in pursuit of deer. Intriguingly, in the early Neolithic, when that bow broke, yew did not grow in Scotland. Either the bow or the bow stave had been taken to Scotland deliberately, possibly from Cumbria or perhaps even Ireland. Yew was clearly already a prized timber for bow-making. Known as the Rotten Bottom bow, it is now housed in the National Museum of Scotland, where it has been honoured by a modern reconstruction to illustrate what a beautiful object it once was. We can only guess at the many other beautiful bows vanished in the dusts of time. Leaving the bow on that peaty hillside had another unforeseeable result: today, the Carrifran Wildwood project has drawn inspiration from that Neolithic bow as they work to restore the area local to where the bow was found to the virgin forest its owner would have traversed on the day the weapon broke.

Archery is a beguiling pursuit that links us with the spirits of our ancestors but also to the forest and the trees. While it has faded from modern lives, just occasionally we are reminded of its former importance. In 2010, the Reverend Mary Edwards, of Collingbourne Ducis near Marlborough, invoked an unrepealed law requiring her parishioners to come to the recreation ground and practise archery. The purpose was to celebrate the building of a new lavatory in the church, and the attendees were treated to a barbecue and live music. What might Henry VIII have thought?

Wild Cherry

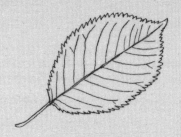

Prunus avium means 'plum of the birds', which is just what this is. Song thrushes and blackbirds gobble the cherries and drop their seeds as they go, unknowingly guaranteeing themselves another cherry crop. Seedlings take just a couple of weeks to emerge from the soil but, after 10 or 20 years, the tree still won't be fully grown. It will, however, begin to produce fruit in its fourth year. With three- to four-year lifespans, most of these birds will not be around for harvest, but their offspring might be.

The wild cherry's lifespan is more substantial at 60 years, and it grows as high as 30 metres. It is widely recognised for its glossy bark, a reddish-brown, reminiscent of a dark honey, with pale lenticels encircling the trunk. The strong, hard timber is popular for furniture and veneers; a good polish secures an impressive shine. It also burns well, giving off a pleasantly sweet-scented woodsmoke.

Buds with overlapping scales cluster on grey branches. Both flowers and leaves can grow from these buds, with a distinctive pair of red glands on the stalk at the base of each serrated ovoid leaf. This foliage is sustenance for caterpillars of many moth varieties.

With an aroma akin to that of the tree's woodsmoke, wild cherry's five-petalled white flowers are cupulate and grow in bunches of up to six. Flowering in spring, wild cherry offers early availability of pollen for bees. Once the bees have done their work, the flowers begin to transform into spherical dark-red cherries. These are food for badgers, dormice and more.

A wild cherry sapling can be carved into a brush, and the bark can be used to fashion baskets similar to birch bark. A green sapling bored through and fitted with a seasoned base makes a beautiful coppery shrink box.

Wild Service

The Romans used the fruit of *Sorbus domestica* to flavour beer, and it's from a slight mutation of the Latin for 'beer', *cervisia*, that the tree got its name: service. The fruit of the related wild service tree (*Sorbus torminalis*) was pressed into similar service, flavouring not just beer but liqueurs, whiskeys, jams and preserves in medieval England. By the nineteenth century, these berry-like fruits, actually pomes with a taste similar to that of dates, had acquired the name 'chequers', and children ate them as sweets. In the twentieth century, they fell from favour; chequers disappeared from market stalls, and the wild service tree became rare.

Happiest in lime-based and clay soils, wild service is native to the UK and is most likely to be found in oak or ash woods in the south and southeast of England. It will grow to 20-25 metres and has dark-brown to pale-grey bark, chequered with cracked squares and becoming scaly as it ages. Its small, glossy, pealike buds form on short leaf stalks and unfurl into lobed leaves, not unlike the Norway maple.

White flowers form in clusters in May. By July, the polli-nated flowers have formed fruit clusters; by August, they are green-brown ovals which, as they ripen through the autumn, become mottled with small, pale spots. These chequers need bletting to make them edible, although this can occur natu-rally if there are sufficient autumn frosts, after which they are brown-speckled.

Wild service trees supply a fine-grained, pliable hardwood that was once used to make billiard cues and musical instruments. Now rare, it is only used for decorative veneers. Service wood is also a good bow wood.

Birch sap is a traditional refreshing tonic that comes direct from the forest and provides nutrients following the passing of winter's depths. Drinking birch sap, we feel we have become one with the forest. While it is definitely enchanting to tap it in the wild, today there is no need to tap a tree and risk harming it, as birch sap can easily be bought from commercial suppliers.

Charred hazelnut remains are a common feature of Mesolithic dwelling places. A staple food of the past, they were cooked underground in a shallow ground oven with a quick, hot fire. Cooking transforms them from a dry nut to a food more like a potato or cooked chestnut.

The Fortingall Yew is possibly the oldest living tree in Britain and Europe. Believed to have seeded sometime in the Bronze Age, it could possibly be as old as five thousand years. Once it was a massive tree that in 1769 had a girth of 17m, but the trunk has suffered through the ages, being broken and damaged by souvenir hunters and fire. Today, protected by a stone wall, the two remaining fragments of the original bole continue to grow vigorously, supporting the reputation of yews as eternal trees.

Of Britain's 13% tree cover, 50% is represented by conifers, mostly plantation blocks established to meet timber demands following the two World Wars. While these woodlands play their part, they have poor species diversity and provide only poor habitat for wildlife when compared to natural conifer or broadleaved woodland. Gradually efforts are being made to restore our natural woodland.

This is the West Dart River valley looking north from above Two Bridges. In the distance the tiny Wistman's Wood clings to the east side of the valley. Riparian valleys such as these present potential opportunities for natural regeneration by reducing grazing pressure. Where such schemes have been implemented, they not only buffer existing woodland and provide new habitat and increased biodiversity but can also form wildlife corridors that link otherwise isolated habitats.

Indicators of Ancient Woodland

Ancient woodland cannot be replanted, so the first step towards the restoration of our ancient woodlands is to recognise them before they are destroyed. These are some of the plant species that can be used to recognise ancient woodland.

Bluebell

Primrose

Pendulous sedge

Toothwort

Ramsons

Yellow archangel

Hart's tongue fern

Wood anemone

Ancient oak

Willow

WHITE WILLOW

The white willow (*Salix alba*) is native to Asia, North Africa, Europe and Britain. Here, it is mainly found in the south of England, most likely in low valleys or alongside rivers and streams, its preferred location being in wet ground. It is the largest of all the willow species, reaching as high as 25 metres when fully mature. Its bark and supple, slender twigs are grey-brown; the ageing bark slowly becomes deeply fissured.

Some trees earn their common names very literally, and white willow does look intensely white each summer. Its shoots and green buds are covered in white hairs, and so are its pale and narrow oval leaves. Both sides of the young leaves are hairy, giving them a silverish look as they hang from the branches; as the leaves grow older, the long, white hairs become sparser on top, but the undersides retain their hairiness.

As if the combined pallor of white willow's bark, twigs, shoots, buds and leaves were not enough, its long, yellow male catkins appear in spring, attracting the insects that will pollinate the slightly smaller, pale-green female catkins hanging from sepa-rate nearby trees, between March and June. After pollination, the females lengthen and grow small pods of minuscule seeds entirely covered in a downy white fluff that will make them extra-buoyant when they're released into the breeze. Until then, the white willow has become a little whiter.

While chaffinches, goldfinches, greenfinches, willow tits and more nest in the willow's branches, caterpillars head for the leaves. Moth species whose caterpillars feast here include the eyed hawk-moth, the puss moth, the red underwing and the willow ermine. And while the caterpillars pick at the leaves, the birds pick off the caterpillars.

Salix alba var. *caerulea* is a white willow variety whose wood is used to make cricket bats, hence its common name, cricket bat willow, although bat-makers tend to call it English willow for marketing purposes. Its timber is lightweight but tough, flexible and unlikely to split, so it is the ideal substance for something that's going to be used exclusively for smacking red-leather spheres with twine-bound cork interiors. This willow is also an essential component of the classic East Sussex basket, the trug.

GREY WILLOW

The grey willow (*Salix cinerea*), also known as pussy willow, is a shrub native to Britain, Europe and western Asia. It thrives in hedgerows and woodland, and it grows near rivers, streams and canals as it prefers damp environments. Fully mature, it will grow no higher than ten metres. As it ages, diamond-shaped fissures begin to appear in its grey-brown bark, and its hairy twigs lose their hairs and become smooth. The shoots, too, have a hairy covering. Small, red buds hug close to the shoots.

Most willow species' leaves are long and thin. The grey willow's leaves are ovate, but they are twice as long as they are wide. These full-sized leaves may obscure the miniature leaves nestling at their base around the shoots. Like the twigs and shoots, the grey willow's young leaves start life hairily, with a layer of fuzzy silver hairs below and reddish hairs beneath the veins. Ageing removes the hairs from their top sides and leaves the undersides only patchily hairy.

The beginning of spring heralds the emergence of grey willow's male and female flowers. One tree bears thickset oval catkins, the grey male that will ripen to yellow; a separate tree has the longer, green females. Both types are fuzzy, but only the wind-pollinated females will turn into the plant's light and fluffy seeds. Like some other willows, the grey is not reliant on wind or insects to carry its seeds; it can self-propagate. It sinks its branches groundward where they grow fresh roots. For this reason, we actually find it easier to grow grey willow from a cutting than from its seeds: the grey willow is already adept at cutting out the middle man with the secateurs.

The bark of every willow species, and of several other plants, produces a bitter-tasting chemical compound with the *Salix*-derived name of salicin. There is a 2,400-year-old history of willow being used in pain relief, and German scientists were maintaining this tradition when they used salicin in the first synthesis of aspirin in 1899. When medieval Europeans chewed willow bark to relieve toothache, they were unknowingly dosing up on salicin.

GOAT WILLOW

The goat willow (*Salix caprea*) is also known as pussy willow and, like the grey willow, is native to Britain, Europe and Asia. It flourishes in much the same habitats as the grey and also does not grow beyond ten metres. The interwoven ridges on its grey-brown bark distinguish it, as do its leaves – these are ovate and hairy, like those of the grey willow, but their tips give the odd impression that someone has been mischievously bending or twisting them out of true. Outsized yellow-brown goat willow buds protrude from the shoots.

The male flowers appear towards the end of March, before the leaves unfurl. These are catkins again, and will turn yellow once they are full of pollen in April, when the longer females, on separate trees, are green. The males are popularly thought to look a lot like cats' paws. Come May, the female catkins grow into fluffy seeds, which are chiefly dispersed on the wind.

Goat willow caters for many types of wildlife, not least a purple royal rarity, the emperor butterfly, for which the tree is the main foodplant. Dusky clearwing, lunar hornet clearwing, sallow clearwing and sallow kitten are among the moth species whose caterpillars ravage the goat willow's foliage.

Mature goat willow trees can live for 300 years, and their timber is yellow and soft and makes a good firewood. The wood and twigs of most willows are sufficiently flexible to be used in crafts like basket-making, but goat willow's twigs are unsuitable for weaving because they are very brittle.

As a *Salix*, salicin is produced by goat willow's bark, which may be boiled in water to extract a multipurpose liquor. Drink it and enjoy respite from back pain, diarrhoea and arthritic joint inflammations; gargle it to relieve a sore throat; apply it to clean an open wound and stop bleeding. Many herbalists recommend it for treatment of plenty more, including bursitis (housemaid's knee), fever, flu, menstrual cramps and tendonitis. Its possible side effects are reportedly mild, if daunting, reading like the lawyers' lists that have replaced the instructions folded into pharmaceutical products' packets: stomach upsets, ulcers, nausea, vomiting, skin rashes, tinnitus ... Small wonder some willows weep.

• • •

Willows are the most ubiquitous friction firewood, white willow being the best of the bunch. The inner bark was an important source of cordage, ideal for twisting into string or strong rope. The bark was also used to tan leather.

Yew

When it turns 900, a yew tree finally qualifies as ancient. In Britain alone, there are ten of them that almost certainly predate the tenth century, and almost a thousand that are more than 500 years old. It is fair to say that yews are a long-lived species. They have a potential lifespan of 3,000 years, so that 'ancient' 900-year-old is really fresh out of college and trying to work out what to do with its life. How much life must a 900-year-old yew have witnessed and sustained in all that time?

The common yew (*Taxus baccata*) is an evergreen conifer, native to Europe, Turkey and Iran. It is a British native, too, mainly growing in the south, although our oldest living example is the Fortingall Yew, which put down roots in a Scottish village in Perthshire some 2,000, 3,000 or perhaps even 5,000 years ago. Left to grow naturally, yews will appear on chalk downs and limestones or in oak or beech woodlands. We have a centuries-long history of planting them in churchyards, though nobody really knows why. They are also popular additions to parks and gardens and often used as a hedging plant.

Yews can grow as high as 20 metres, although they are exceptionally slow-growing so it will take many years. The tree's peeling bark is scaly and reddish-brown with hints of purple that increase with age. As an evergreen, yew holds on to its leaves all year round, two rows of small needles, one on each side of every twig. These needles have dark-green top sides, but the undersides are a grey-green, and every one culminates in a sharply pointed tip. The leaves remain green for three years before eventually turning brown.

The yew blossoms between February and April. It is dioecious, so some trees will be harbouring the unimposing yellow-white sacs of small spheres that are the males, while others will be displaying scaly green females that look like buds. The males appear first, and reveal themselves as clusters of stamens, rather than true flowers. When the females emerge, they also turn out not to be actual flowers. They are ovules, containing all that's truly needed: the germ cell awaiting pollination. Over the summer, each will transition into an aril, a single seed at the heart of what looks like a fleshy red berry, open at its tip.

The yew is classed as a conifer because it grows seed-cones. But they don't look like cones, they look like extremely enticing berries, yet they are not berries. What they are, perhaps, is a trap: at least part of achieving longevity bordering on immortality is devising ways to despatch and deter assailants, and any aggressor that makes it past the bright red, sweet and tasty fruit flesh to the seed may soon stop being a prospective threat. Though the aril is not, the seed inside it is poisonous, containing taxine alkaloids that are toxic to humans and most animals. There are a few who can cope: it passes straight through most birds, dropping to the ground inside a ready-made fertiliser; and a badger appears to enjoy the fruit without digesting the tiny seeds at all, expelling them intact with the rest of its waste.

· · ·

Among the oldest extant wooden artefacts is the spearhead that was excavated in Clacton, Essex, in 1911. Its dating to roughly 420,000 years ago means that it is the oldest-known tool made of worked wood. The wood, inevitably, is yew, chosen for its springiness and ability to retain a sharp point.

Two factors have, luckily, got in the way of yew becoming the staple of industrialised production that it might have been. First, there was for a time huge demand for this tough timber, which led to heavy felling and, thus, a dramatic decline in numbers; something that takes decades to grow can't be replaced overnight. Second, it is a timber that demands very wary treatment. Recall for a moment that red 'berry' flesh and its lack of toxicity. What confirms it as a trap, or bait, is that every other part of a yew tree is poisonous. Incautious handling can have unpleasant results: inhalation of yew sawdust can trigger breathing difficulties, sneezing fits, headaches, giddiness, irregular heartbeat and dermatitis; there are some very unpleasant reports of oedema cases brought on by reaction to yew. It does not impact on everybody by any means, or even most, but it does affect more people than any other wood.

Yew's noxiousness offers one explanation for the phenomenon of the tree's extensive presence in Britain's churchyards. The theory is that a yew tree's risks to their livestock discouraged the local peasantry from allowing their cattle to graze on church ground, inflicting damage and defilement. It is certainly possible; modern governments still protest faith in the power of deterrents, so why not medieval clergy?

There is also a chance that this is a case of the early Christian church absorbing and adapting key elements of pre-existing religions in order to smooth a populace's transfer from pagan beliefs to the Faith. Our knowledge of authentic druidic history is fragmentary and corrupted, but it seems the yew was considered sacred by Celtic druids. Perhaps a concerted effort to subsume some local paganism explains the presence of 500 or so church-yard yews that are older than the actual churches.

It seems that the yew tree has long been a symbol of both death and life, and not just for the original Celtic druids. Ancient Greek mythology associates the yew with the goddess Hecate, liberator of souls after death. Worshippers used to be encouraged to carry yew branches to church on Palm Sunday. One name for it means Tree of the Dead. Another reason for this association may be that the heartwood of yew is red and when wet can seem to bleed; blood is present at birth and sometimes death.

Intriguingly this toxic tree that once supplied us with deadly weaponry is today being harvested to produce medicine to combat soft-tissue cancers. Truly it is a tree of both death and life.

10

THE FUTURE

Nuthatches are one of the species benefitting from global
warming, year on year extending their presence northwards.

t is iron and, later, steel tools, that have most dramatically changed our relationship with our woodland. The Romans intensified both our agriculture and our iron production. By the time they departed our shores, iron tools were widely available. Now every tree species could be easily felled and worked, no matter its size or strength. Even dry, seasoned wood, which is much harder to carve than green (fresh-cut) wood, could now be easily worked.

The stage was set for the emergence of modern woodworking. The first obvious change to occur was house design. The traditional round houses, constructed from coppice poles and hazel wattle, went out of fashion, to be replaced by rectangular frame houses made from squared beams of more durable oak. The carpenter with felling axe, saw, side axe, chisel, drill and adze would construct the house frame from green timber. In the woodland which supplied the timber, the trees were felled, cut to size, squared, jointed and assembled. Then, in a forerunner of today's kit houses, the frame would be disassembled and re-erected where the house was to be built.

The growing ambition of our farming communities was also realised in our shipbuilding. The Sutton Hoo ship burial attests to both the skill of Anglo-Saxon shipwrights and the increased trading communication during the Dark Ages. Britain's agricultural wealth would attract the covetous eye of the Vikings who, after brutally sacking north-eastern coastal abbeys for their ecclesiastical ornaments in the eighth century, invaded en masse in the ninth century to appropriate the superb agricultural land of northern Britain for themselves.

Norse ambitions would finally surmount Anglo-Saxon tenure of the land with the Norman invasion of 1066. Twenty years later, William I ordered a Great Survey to be conducted. It is recorded in the manuscript we know today as the Domesday Book. This was the first reliable, detailed, systematic measure of land usage in England and much of Wales. From it, we learn that woodland was valued as grazing pasture for swine, for the supply of fuel and for salt-making. The trees now most valued were oak, alder, osier and ash. Woodland was already mostly confined to the clay soils less suited to agriculture, and is estimated to have represented around 15 per cent of the land area surveyed. In the following centuries, under the demand of the increasing population, it is believed that woodland was further reduced to an astonishingly low seven per cent.

Then, in 1348, a seaman returning from Gascony landed in Weymouth, Dorset. He was infected with *Yersinia pestis*, better known as Bubonic Plague. This disease spread like a forest fire through the human population. By autumn of the same year, the crowded streets of London were infested with the Great Pestilence that today we call the Black Death. By the summer of 1349, the whole country was affected, yet the disease seems to have burned itself out by December of the same year.

The damage was staggering: a thousand villages had been lost; recent estimates put the death toll at between 40 and 60 per cent of the population. Over the next century there would be several more outbreaks, most notably in 1361–62, when a further 20 per cent of the population was claimed. The consequences of the disease were far-reaching. The contemporary French chronicler Froissart estimated that a third of the population of Europe had perished. It brought a halt to the Hundred Years' War and would cause a labour shortage with accompanying social unrest, including the Peasants'

Revolt. For our woodland, though, the reduced population offered a timely reprieve. With fewer fieldworkers available, arable farming reduced and pastoral land use increased, with a resulting growth in our forest cover.

The fourteenth century would also see advances in woodworking practices. While in small villages the carpenter needed to be a jack-of-all-woodworking-trades, in larger cities a woodworker would become a specialist: carpenter, sawyer, joiner, turner, carver, box-maker, coffer-maker, bowyer, fletcher, upholder or wheelwright. In many cases, this required the undertaking of a seven-year apprenticeship. Trade organisations emerged by which associated woodworkers agreed to abide by a professional code of conduct. This regulation ensured high standards, could offer the woodworker a degree of income security and provided customers with quality assurance.

Although they rarely survive, common everyday items were fashioned with a rustic woodworking skill that was commonplace in medieval Britain. Many fine examples of this handcraft, made from coppice woods, can be seen in the extraordinary finds salvaged from the wreck of the *Mary Rose*. But Britain's overall demand for timber could not be met by domestic supply alone. High-quality timber was already in short supply. Oak, for example, had been imported from the Baltic since the middle of the thirteenth century.

Expensive materials could therefore not be wasted. It would be this shortage of wood that would influence refinement in all trades of British woodworking. Wheels would become more sophisticated, furniture more elegant and decorative masterpieces would be carved in wood, such as the lime panels carved by Grinling Gibbons, one of the greatest woodcarvers to have ever lived. Perhaps most famous of all was Thomas Chippendale, whose useful and ornamental designs epitomised the zenith of eighteenth-century cabinet-making. While embracing the taste and fashion of his time, Chippendale's work transcended mere mastery of tools and technical skill. He was a cabinetmaker who saw his craft as a science, drawing inspiration from classical architecture and executing his pieces with refined perfection. Despite the deliberate and obvious precision of his work, his furniture was also soulful, demonstrating a warmth and feeling for the wood itself, a desire to share his passion for the raw material by exhibiting its properties in designs that could not be ignored. His genius has influenced cabinet-making ever since. In 1754, he published *The Gentleman and Cabinet-Makers Director*, a book on cabinet-making and the first design catalogue of its kind. Woodworking had entered the modern age.

So where are we today?

We have never had better information about our woodlands than we have today, and a quick dive into the statistics is a sobering experience. In 1905, the UK had less than five per cent woodland cover. Fortunately, over the past hundred years it has more than doubled, today covering 3.2 million hectares of land, or 13.2 per cent of the land surface. Only around half of that is native woodland, however, with the majority of the rest comprising commercial plantations of non-native species that offer poor habit opportunities for wildlife.

The good news is that Britain is one of the few countries where woodland coverage is increasing. Trees outside of woodlands – in hedgerows, on field margins, in parks and city streets – cover just over three per cent of the land. These are important trees. They provide wildlife with connectivity between woodland habitats. Only poor data exists for trends here, although at least one regional study has recorded a loss of half such trees since 1850. In my own lifetime, I have witnessed trees vanish entirely from town streets and urban neighbourhoods. In one area that I know well, open skies now replace the forest canopy in which I last watched a wryneck in Britain. It is disappointing that, despite our increasing understanding of woodland, its loss continues to be of concern. This is never more the case than for our ancient woodland.

I remember the shock of listening to a local councillor answering questions about his controversial road development, which at the time was threatening and would later destroy an ancient woodland. 'I don't know what all the fuss is about,' he said. 'We can plant a new ancient woodland nearby.' His crass statement immediately revealed his ignorance of the issue and that he was not qualified to make the decisions he was entrusted with. It shouldn't need saying, but ancient woodland is special *because of its age*. It cannot be planted; it must be generated naturally and takes many centuries for its unique complex of organisms to develop.

Ancient woodland now accounts for only 2.5 per cent of our land area and in many areas is still seriously threatened by housing development, roadbuilding and rail expansion. When we consider constructing new roads, which I acknowledge is sometimes necessary, we should remember that we are discussing the destruction of long-established ecosystems to support the passage of automobiles, an invention that is not yet as old as a veteran oak tree.

The UK is, however, rich in ancient and veteran trees, with over 120,000 recorded and probably more yet to be so. Such trees provide unique ecosystems and a living link to our archaeological past. However, while such trees are recognised in planning guidance, and some individual trees have special protection, the majority have no protection in law at all. The good news is that the threats that most ancient trees face are posed by overgrazing, overshading or soil compaction by the feet of too many admiring visitors. All are easily remedied problems.

So what about the health of our woodland? For a wood to be considered ecologically healthy, it needs to exhibit a diversity of tree species and tree ages. There should be open glades within the woodland, fallen deadwood, diverse forest flora and veteran or ancient trees – in other words, the conditions which once naturally occurred in our original woodlands. Today, only seven per cent of British woodland is classified as being ecologically healthy. This should be our greatest concern. Our woodland is facing a plethora of new threats, from climate change, pollution and invasive species.

Climate change is quantifiable in British woodland, where spring now begins eight days earlier than it did a hundred years ago. This may not seem much, but the forest ecosystem is finely attuned to the natural calendar. Such a change can have catastrophic consequences for woodland birds that time their brood hatch to the availability of specific caterpillars. It may also be a contributing factor in the sharp decline recorded for populations of woodland birds, moths and butterflies over the past 50 years. Of course, there will always be some species that can exploit such changes, like the nuthatch, whose whistling heralds the arrival of spring, a sound that is moving northwards with its expanding territorial range.

More insidious and difficult to observe is the effect of nitrogen fertilisers from agriculture that are destroying the forests' flora of lichens and unbalancing the ecology in ways that are yet to be fully understood. There is much more that could be said but, honestly, it is overwhelmingly bleak news. Our forests are in crisis and desperately need our help.

But, looking for a glimpse of hope, woodland is very much on the agenda. There are ambitious plans in place to increase woodland coverage in the UK. And perhaps more importantly, our children are being taught at school to appreciate and value woodland.

Apart from the many traditional woodland products we have historically enjoyed, trees have much to offer us. As well as providing oxygen during the day, a mature beech tree may provide enough oxygen each year to support ten people. Trees also sequester carbon from the environment, locking it away in their tissues. This has been a driving factor in recent tree-planting schemes, as the country strives to become carbon neutral by 2050. To achieve those targets, however, we will need considerably more trees. It is worth noting that while ancient woodland represents 25 per cent of British woodland, it holds 36 per cent of our woodland carbon. Looking towards the future, we should look for opportunities for natural woodland to fully mature into ancient woodland and encourage natural woodland regeneration wherever possible, to provide a legacy of ancient woodland for our far-distant descendants.

There are many other ways in which trees benefit us. They are masters, for example, at slowing the runoff of water from the land. This can reduce or prevent downstream flooding and soil erosion. Equally importantly, it creates damp ground, increasing biodiversity. As we should all know, this is the first real rule of nature: where there is water, there is life.

In our towns and cities, trees significantly filter the air of dust and pollutants and provide welcome shade, enhanced by the transpiration that further cools the temperature beneath them. Where they have been lost, city trees should be replaced. Access to trees is vital to human wellbeing. Not long ago, I encountered a new pay-and-display car parking scheme in a popular public woodland area. Instead of being busy with people, the car parks were deserted. However they are presented, such schemes are counter to the needs of society, commoditising access to nature for the financial benefit of a minority.

I was recently asked to support a rather unusual promotional scheme. It involved rewilding Trafalgar Square. Liking the idea, I agreed without ever anticipating the power of the experience. In early March 2022, overnight, 3,000 plants and trees were installed at the foot of Nelson's Column. In the dawn light, we waited to see the response of the public and, as the first rays of the sun warmed what was a very chilly flower, a bee alighted in search of nectar. It was the first visitor and a portent of what was to come.

As the day progressed, that small green space attracted ever-increasing numbers of visitors from far and wide. Some simply stared in silence, some were ebullient, others sat and silently breathed in the atmosphere. Everyone was smiling. The rest of Trafalgar Square stood virtually empty that day. It was a real 'happening'. The demonstrable power of nature to move people was awe-inspiring.

A government minister visiting the site told me, 'We don't really like the term rewilding.' Even Lord Nelson, who insisted oak trees were planted to ensure a supply of timber for the Royal Navy of the future, looked down from his lofty perch in dismay. What a foolish comment. We were surrounded by people who absolutely believed in rewilding: the restoration of our natural environment, the reclamation of land outside of farming use to increase biodiversity. In my work with nature, I have learned that acting to increase biodiversity should be our prime directive. Rewilding as a concept has been taken up by the nation. Why not win some votes by steering and supporting the process? I was glad that I was wearing green that day rather than a suit. Sadly, like a fairy tale, the greenery had to be removed and the venerable site returned to its normal state before midnight that same day. But those of us who were there will forever remember the powerful emotions of the day.

Here is the point. Woodland is actually good for us. In 1982, the Japanese Ministry for Agriculture was searching for a way to alleviate growing levels of work-induced stress and a high incidence of auto-immune disease. Confident that if people could be encouraged to invest time in woodland it would help, they looked to ancient Shinto and Buddhist practices and invented what they called *Shinrin-yoku*, which translates as 'forest bathing'. The results were astonishingly successful and attracted the attention of a scientific community keen to explain, prove or disprove the phenomenon.

The basic concept of forest bathing is to take a short leisurely visit to a forest and to wander at will, freeing the subconscious mind to discover small details for itself, without conscious direction. As far as possible investigating the world with all the senses, drinking in the shapes and tones, becoming mindful to the sounds of birdsong or the dripping of rain from leaves. Exploring textures by touch. It sounds very contrived, but the concept is simple. Take time out from the normal hectic pace of life to allow nature back in. Do not imagine that this is just about a change of pace and scenery; when we are amongst trees, they are having a far more profound influence on us than we realise.

Scientific investigation has not only validated the concept, but it has also unravelled some of the astonishing ways in which *Shinrin-yoku* works. Forest bathing enhances our mood, increases our positive feelings and reduces feelings of negativity. It also improves sleep patterns. But more than this, it has revealed that time in woodland boosts self-esteem, encourages calmness and results in improved patience.

Compared to modern life, where we actively direct our attention to the world around us, our attention is grabbed in a more random way by sensory stimulus that intrigues our curiosity in the complex environment of the forest. This provides relief to our direct attention that is overworked by the demands of technological concerns. By that means, we can improve our ability to focus our attention when it is really needed. Children with ADHD have shown marked benefits from regular exposure to woodland and learning in an outdoor classroom. While it is not claimed to be a cure, it has been shown to alleviate anxiety for people coping with stress induced by trauma.

It has also been determined that we benefit physiologically from walking amongst trees. Our pulse rate slows, our blood pressure drops, our levels of cortisol, dopamine and blood glucose stabilise. Most intriguing of all, we are exposed to phytochemicals that boost our immune system and increase our natural killer cells that enable us to fight off disease. Enjoying woodland is thus a preventative medication. It is a form of natural aromatherapy that is freely available, has no known side effects and requires no prescription. And do the effects last? Here, Japanese researchers have shown that even short exposure of a few hours to the forest can last for seven days.

Perhaps we should not be surprised to discover these things. After all, we all have a hunter-gatherer inside us. The forest was and still is our natural home. It is only in the past few thousand years that we have turned our axes and saws to the trees that once provided us with sanctuary. As we face the need to halt and correct centuries of environmental abuse, I wonder, can we afford to live without trees? One US research project monitored the health of a community losing their trees to disease. The results indicated an associated increase in cardiovascular and lower pulmonary illness.

When I was at Trafalgar Square, a radio interviewer asked me, 'So Ray, do you hug trees?' My answer was 'Yes. Don't you? After all they are sentient, they sense sunlight and, like you and me, they are imbued with the vital spark that is life itself.' Walking in woodland, I genuinely feel that I am surrounded by friends.

Today, as we stand in the Sub-Atlantic phase of the Holocene, we are witnesses to a blossoming of human ingenuity in science, technology, civilisation, medicine and communication. When I gaze out into space and wonder what is out there, I am just like those Palaeolithic communities that once gazed at wilderness Britain. I find myself reflecting that it is also in the last 2,500 years that we have seen global wars, anthropogenic environmental modification, pollution, species loss and climate change on a scale that now stands to rival the influence of our planet's sun. Ironically, we now have more choice in our destiny than ever before. The question is whether we have the collective wisdom and political will to make the wise choices. We have tended always to think of what trees can do for us. Why don't we instead ask what we can do for our trees?

We can make a start by protecting our ancient woodland and veteran trees; by recognising their unique role in the larger environment. To do so we shall have to study the soil and understand the way pollution damages the essential mycorrhizal network, that is so deeply rooted in the lives of our oldest trees. We must, of course, do more to establish greater woodland cover. This is happening in a drive to sequester carbon, but it will also greatly benefit biodiversity, helping to slow the alarming pace of species loss while improving our lives generally.

Despite the monumental problems we face, I am filled with hope. The coming generations are and will be better informed about environmental issues and are clearly prepared to act for change. Our methods of scientific investigation have greatly advanced, as have our means to communicate globally about local issues. If we get it right, future generations will be able to walk through forest glades, rich in a multitude of diverse life forms, a world perhaps last witnessed by our hunter-gatherers. If so, they will certainly pause a moment in contemplation and say, 'Isn't it beautiful. I love British woodland.'

ACKNOWLEDGEMENTS

Trying to visualise the lives of our ancestors, for whom our trees were such a vital resource for life, is not easy. In this exploration of our woodland, I have come to realise that the people of Britain seem to have always been swift adopters and developers of new ideas and technologies. Being first at the remote, northwest corner of Europe during the frigid years following the last ice age, and then separated from mainland Europe when Doggerland vanished under rising sea levels, it may have been our geographical location that shaped our attitude and boundless resourcefulness. Helping me to understand this and to visualise the world of our ancestors has been an army of silent partners, including many extraordinary archaeologists. The quality of their investigations is astonishing. Such work is taking on increasing significance as we study the effects of climate change in the past to understand the challenges of our future. Of special note has been Professor Nicky Milner from the University of York, without whose expert guidance I would have become totally turned around on the trail of our Mesolithic and Late Upper Palaeolithic ancestors. Thank you, Nicky, for your support and for sharing the treasure trove of information that has been unearthed at Star Carr.

At the Mary Rose Trust, I must thank Dr Alexzandra Hildred and Sally Tyrell for being so generous with the time they afforded me to examine and photograph the longbows salvaged from the wreck. Your commitment and enthusiasm are uplifting and honour the many sailors who perished aboard the ship.

Also, and importantly, I thank my tutors from the many First Nations who I have had the opportunity to learn from. I appreciate that maintaining the traditions and culture of pre-agricultural lifeways is increasingly challenging in today's technologically interconnected world. Yet I am more certain than ever, that in the wisdom and attitudes that you safeguard, are many solutions to the environmental and societal problems that humanity will face in the coming century.

At Ebury I must thank Lorna Russell for initiating this project and Claire Collins for bringing together all the many components so ably. You have both been a joy to work with. I would also like to mention Steve Tribe.

As always, my thanks go to my agent, Jackie Gill. Thank you for working as you do to manage business in a friendly and harmonious way. It means so much to me.

Lastly and most importantly of all, thank you to you dear reader. Your support and curiosity are the source of my hope for the future of British woodland, and in fact, for woodland worldwide. Together I know that we can rescue, protect and restore a wonderful sylvan legacy to pass on to future generations.

INDEX

Note: page numbers in **bold** refer to information contained in captions.

Mesolithic peoples 21–2, 68, 100–1, 110,
134–5, 143, 145, 147, 150, 156, 160–1,
168, 182–4, 188–91, **209**, 271
metal on wood 237–44, **237**
midden mounds 53, 133, 184
milk-cap mushroom 123
moss 10, 29, 51, 72, 101, 123, 187
moths 24, 27, 29, 56, 84, 87, 121, 123,
128, 130, 170, 172, 174, 198, 208, 230,
235, 275, 280, 295
mullein 113
musical instruments 57, 127, 175,
177–9, 203, 206
Must Farm Quarry 225, 240
mycelium **47**, 48
mycorrhizal fungi 46–8, **47**, 300

nails 242
narcotics 82
Neanderthals 94–5, 99, 214, 218, 255
necklaces 212–13
Neolithic peoples 26, 100, **181**, 193–5,
238, 244, 272–4
nettle 137, 140, 165, **209**, 215–17, 223–5
New Forest 201
nitrogen fixation 23–4
nomadic lifestyles 183–6
Norman conquest 26, 291
North America 81–2
North Star 72
Northern Ireland 200
Nunataaġmiut (Nunamiut) 186–7, 191
nuthatch 29, 124, **289**, 295
nuts 122–3, 129, 160–5, 168, 189–90

oak (*Quercus*) **21**, 42, 74, 122, 128, 134,
161–2, 177, 182, 198–201, 210, 218–20,
225, 234, 238–9, 290–2, 297
Eagle 201
English (*Q. robur*) 73, 199, 200
pedunculate 199
sessile (*Q. petraea*) 73, 199
Okanagan people 156–7
Ötzi 239–40, 251, 272

oyster mushroom 56

pack animals 8, 185–6, 187–8
Palaeolithic 7, 19–20, 97, 258–9, 261,
300
Late 99, 147, 183, 185, 257
Middle 94
Upper **1**, 100–1, 107, 187
pear 26, 148–9, 153, 202–3, 226
common (*Pyrus communis*) 202
penny bun 47
photosynthesis 24, 45, 73
phototropism 73–6
Phytophthora 24
pignuts 132, 142–3
pine **21**, 42, **67**, 73–6, 139, 157, 165–6,
204–6, 214, 218–20, 272
pioneer species 204
plastic waste 196
plum 26, 145
poisonous species 47, 57, 62, 64, 80, 83,
118, 130, 135–43, 148–51, 155–62,
286–7
pole lathes 243
pollarding 239
pollen 41, 123
pollination, wind 40
pollution 196, 296, 300
poplar 189, 207–8, **207**, 267
black (*Populus nigra*) 208
white (*Populus alba*) 207–8
porcelain fungus 56
pot boilers 53
potter's wheel 243
ptarmigan 38, 39

qulliq (fat lantern) 10–12, **11**

ramsons 142
raspberry, wild 140, 147–8
redcurrant 144
reindeer **1**, 5–7, 16–18, 20, 26, 38, 55,
153, 185–6, 213, 257
resin 206